D1606319

CAMBRIDGE STUDIES IN ADVANCED MATHEMATICS 145

LECTURES ON LYAPUNOV EXPONENTS

The theory of Lyapunov exponents originated over a century ago in the study of the stability of solutions of differential equations. Written by one of the subject's leading authorities, this book is both an account of the classical theory, from a modern view, and an introduction to the significant developments relating the subject to dynamical systems, ergodic theory, mathematical physics and probability. It is based on the author's own graduate course and is reasonably self-contained with an extensive set of exercises provided at the end of each chapter.

This book makes a welcome addition to the literature, serving as a graduate text and a valuable reference for researchers in the field.

Marcelo Viana is Professor of Mathematics at the Instituto Nacional de Matemática Pura e Aplicada (IMPA), Rio de Janeiro.

CAMBRIDGE STUDIES IN ADVANCED MATHEMATICS

All the titles listed below can be obtained from good booksellers or from Cambridge University Press. For a complete series listing visit: www.cambridge.org/mathematics.

Already published

107 K. Kodaira *Complex analysis*
108 T. Ceccherini-Silberstein, F. Scarabotti & F. Tolli *Harmonic analysis on finite groups*
109 H. Geiges *An introduction to contact topology*
110 J. Faraut *Analysis on Lie groups: An introduction*
111 E. Park *Complex topological K-theory*
112 D. W. Stroock *Partial differential equations for probabilists*
113 A. Kirillov, Jr *An introduction to Lie groups and Lie algebras*
114 F. Gesztesy *et al. Soliton equations and their algebro-geometric solutions, II*
115 E. de Faria & W. de Melo *Mathematical tools for one-dimensional dynamics*
116 D. Applebaum *Lévy processes and stochastic calculus (2nd Edition)*
117 T. Szamuely *Galois groups and fundamental groups*
118 G. W. Anderson, A. Guionnet & O. Zeitouni *An introduction to random matrices*
119 C. Perez-Garcia & W. H. Schikhof *Locally convex spaces over non-Archimedean valued fields*
120 P. K. Friz & N. B. Victoir *Multidimensional stochastic processes as rough paths*
121 T. Ceccherini-Silberstein, F. Scarabotti & F. Tolli *Representation theory of the symmetric groups*
122 S. Kalikow & R. McCutcheon *An outline of ergodic theory*
123 G. F. Lawler & V. Limic *Random walk: A modern introduction*
124 K. Lux & H. Pahlings *Representations of groups*
125 K. S. Kedlaya *p-adic differential equations*
126 R. Beals & R. Wong *Special functions*
127 E. de Faria & W. de Melo *Mathematical aspects of quantum field theory*
128 A. Terras *Zeta functions of graphs*
129 D. Goldfeld & J. Hundley *Automorphic representations and L-functions for the general linear group, I*
130 D. Goldfeld & J. Hundley *Automorphic representations and L-functions for the general linear group, II*
131 D. A. Craven *The theory of fusion systems*
132 J. Väänänen *Models and games*
133 G. Malle & D. Testerman *Linear algebraic groups and finite groups of Lie type*
134 P. Li *Geometric analysis*
135 F. Maggi *Sets of finite perimeter and geometric variational problems*
136 M. Brodmann & R. Y. Sharp *Local cohomology (2nd Edition)*
137 C. Muscalu & W. Schlag *Classical and multilinear harmonic analysis, I*
138 C. Muscalu & W. Schlag *Classical and multilinear harmonic analysis, II*
139 B. Helffer *Spectral theory and its applications*
140 R. Pemantle & M. C. Wilson *Analytic combinatorics in several variables*
141 B. Branner & N. Fagella *Quasiconformal surgery in holomorphic dynamics*
142 R. M. Dudley *Uniform central limit theorems (2nd Edition)*
143 T. Leinster *Basic category theory*
144 I. Arzhantsev, U. Derenthal, J. Hausen & A. Laface *Cox rings*
145 M. Viana *Lectures on Lyapunov exponents*

Lectures on Lyapunov Exponents

MARCELO VIANA
Instituto Nacional de Matemática Pura e Aplicada (IMPA),
Rio de Janeiro

CAMBRIDGE
UNIVERSITY PRESS

CAMBRIDGE
UNIVERSITY PRESS

University Printing House, Cambridge CB2 8BS, United Kingdom

Cambridge University Press is part of the University of Cambridge.

It furthers the University's mission by disseminating knowledge in the pursuit of education, learning and research at the highest international levels of excellence.

www.cambridge.org
Information on this title: www.cambridge.org/9781107081734

First published 2014

Printed in the United Kingdom by CPI Group Ltd, Croydon CRO 4YY

A catalogue record for this publication is available from the British Library

Library of Congress Cataloging-in-Publication data
Viana, Marcelo, author.
Lectures on Lyapunov exponents / Marcelo Viana, Instituto Nacional de Matemática Pura e Aplicada (IMPA), Rio de Janeiro.
pages cm. – (Cambridge studies in advanced mathematics ; 145)
Includes bibliographical references and index.
ISBN 978-1-107-08173-4 (Hardback)
1. Lyapunov exponents. I. Title.
QA372.V53 2014
515′.48–dc23 2014021609

ISBN 978-1-107-08173-4 Hardback

To Tania, Miguel and Anita,
for their understanding.

Contents

Preface

1. The study of characteristic exponents originated from the fundamental work of Aleksandr Mikhailovich Lyapunov [85] on the stability of solutions of differential equations. Consider a linear equation

$$\dot{v}(t) = B(t) \cdot v(t) \tag{1}$$

where $B(\cdot)$ is a bounded function from $\mathbb{R}$ to the space of $d \times d$ matrices. By the general theory of differential equations, there exists a so-called *fundamental matrix* A^t, $t \in \mathbb{R}$ such that $v(t) = A^t \cdot v_0$ is the unique solution of (1) with initial condition $v(0) = v_0$. If the *characteristic exponents*

$$\lambda(v) = \limsup_{t \to \infty} \frac{1}{t} \log \|A^t \cdot v\| \tag{2}$$

are negative, for all $v \neq 0$, then the trivial solution $v(t) \equiv 0$ is asymptotically stable, and even exponentially asymptotically stable. The stability theorem of Lyapunov asserts that, under an additional regularity condition, stability remains valid for nonlinear perturbations

$$\dot{w}(t) = B(t) \cdot w(t) + F(t, w) \quad \text{with } \|F(t, w)\| \leq \text{const} \|w\|^{1+\varepsilon}.$$

That is, the trivial solution $w(t) \equiv 0$ is still exponentially asymptotically stable.

The regularity condition of Lyapunov means, essentially, that the limit in (2) does exist, even if one replaces vectors v by l-vectors $v_1 \wedge \cdots \wedge v_l$; that is, elements of the k-exterior power of $\mathbb{R}^d$, for any $0 \leq l \leq d$. This is usually difficult to check in specific situations. But the multiplicative ergodic theorem of Oseledets asserts that Lyapunov regularity holds with full probability, in great generality. In particular, it holds on almost every flow trajectory, relative to any probability measure invariant under the flow.

2. The work of Furstenberg, Kesten, Oseledets, Kingman, Ledrappier, Guivarc'h, Raugi, Gol'dsheid, Margulis and other mathematicians, mostly in the

1960s–80s, built the study of Lyapunov characteristic exponents into a very active research field in its own right, and one with an unusually vast array of interactions with other areas of Mathematics and Physics, such as stochastic processes (random matrices and, more generally, random walks on groups), spectral theory (Schrödinger-type operators) and smooth dynamics (non-uniform hyperbolicity), to mention just a few.

My own involvement with the subject goes back to the late 20th century and was initially motivated by my work with Christian Bonatti and José F. Alves on the ergodic theory of partially hyperbolic diffeomorphisms and, soon afterwards, with Jairo Bochi on the dependence of Lyapunov exponents on the underlying dynamical system. The way these two projects unfolded very much inspired the choice of topics in the present book.

3. A diffeomorphism $f : M \to M$ is called *partially hyperbolic* if there exists a Df-invariant decomposition

$$TM = E^s \oplus E^c \oplus E^u$$

of the tangent bundle such that E^s is uniformly contracted and E^u is uniformly expanded by the derivative Df, whereas the behavior of Df along the *center bundle* E^c lies somewhere in between. It soon became apparent that to improve our understanding of such systems one should try to get a better hold of the behavior of $Df \mid E^c$ and, in particular, of its Lyapunov exponents. In doing this, we turned to the classical linear theory for inspiration.

That program proved to be very fruitful, as much in the linear context (e.g. the proof of the Zorich–Kontsevich conjecture, by Artur Avila and myself) as in the setting of partially hyperbolic dynamics we had in mind originally (e.g the rigidity results by Artur Avila, Amie Wilkinson and myself), and remains very active to date, with important contributions from several mathematicians.

4. Before that, in the early 1980s, Ricardo Mañé came to the surprising conclusion that generic (a residual subset of) volume-preserving C^1 diffeomorphisms on any surface have zero Lyapunov exponents, or else they are globally hyperbolic (Anosov); in fact, the second alternative is possible only if the surface is the torus $\mathbb{T}^2$. This discovery went against the intuition drawn from the classical theory of Furstenberg.

Although Mañé did not write a complete proof of his findings, his approach was successfully completed by Bochi almost two decades later. Moreover, the conclusions were extended to arbitrary dimension, both in the volume-preserving and in the symplectic case, by Bochi and myself.

5. In this monograph I have sought to cover the fundamental aspects of the classical theory (mostly in Chapters 1 through 6), as well as to introduce some of the more recent developments (Chapters 7 through 10).

The text started from a graduate course that I taught at IMPA during the (southern hemisphere) summer term of 2010. The very first draft consisted of lecture notes taken by Carlos Bocker, José Régis Varão and Samuel Feitosa. The unpublished notes [9] and [28], by Artur Avila and Jairo Bochi were important for setting up the first part of the course.

The material was reviewed and expanded later that year, in my seminar, with the help of graduate students and post-docs of IMPA's Dynamics group. I taught the course again in early 2014, and I took that occasion to add some proofs, to reorganize the exercises and to include historic notes in each of the chapters. Chapter 10 was completely rewritten and this preface was also much expanded.

6. The diagram below describes the logical connections between the ten chapters. The first two form an introductory cycle. In Chapter 1 we offer a glimpse of what is going to come by stating three main results, whose proofs will appear, respectively, in Chapters 3, 6 and 10. In Chapter 2 we introduce the notion of linear cocycle, upon which is built the rest of the text. We examine more closely the particular case of hyperbolic cocycles, especially in dimension 2, as this will be useful in Chapter 9.

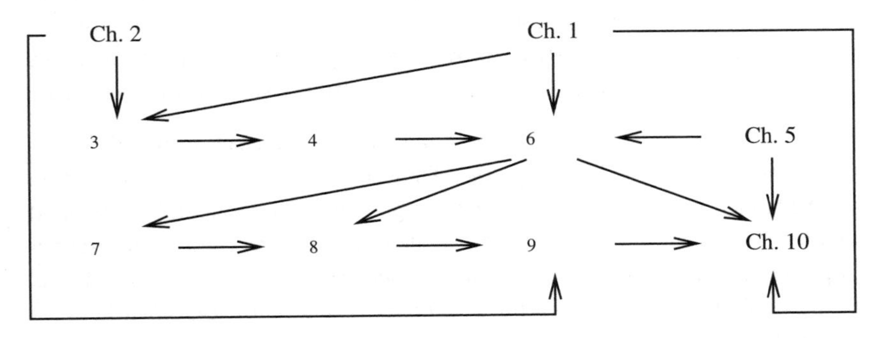

In the next four chapters we present the main classical results, including the Furstenberg–Kesten theorem and the subadditive ergodic theorem of Kingman (Chapter 3), the multiplicative ergodic theorem of Oseledets (Chapter 4), Ledrappier's exponent representation theorem, Furstenberg's formula for exponents of irreducible cocycles and Furstenberg's simplicity theorem in dimension 2 (Chapter 6). The proof of the multiplicative ergodic theorem is based on the subadditive ergodic theorem and also heralds the connection between Lyapunov exponents and invariant/stationary measures that lies at the heart of the results in Chapter 6. In Chapter 5 we provide general tools to develop that connection, in both the invertible and the non-invertible case.

7. The last four chapters are devoted to more advanced material. The main goal there is to provide a friendly introduction to the existing research literature. Thus, the emphasis is on transparency rather than generality or completeness. This means that, as a rule, we choose to state the results in the simplest possible (yet relevant) setting, with suitable references given for stronger statements.

Chapter 7 introduces the invariance principle and exploits some of its consequences, in the context of locally constant linear cocycles. This includes Furstenberg's criterion for $\lambda_- = \lambda_+$, that extends Furstenberg's simplicity theorem to arbitrary dimension. The invariance principle has been used recently to analyze much more general dynamical systems, linear and nonlinear, whose Lyapunov exponents vanish. A finer extension of Furstenberg's theorem appears in Chapter 8, where we present a criterion for simplicity of the whole Lyapunov spectrum.

Then, in Chapter 9, we turn our attention to the contrasting Mañé–Bochi phenomenon of systems whose Lyapunov spectra are generically *not* simple. We prove an instance of the Mañé–Bochi theorem, for continuous linear cocycles. Moreover, we explain how those methods can be adapted to construct examples of discontinuous dependence of Lyapunov exponents on the cocycle, even in the Hölder-continuous category. Having raised the issue of (dis)continuity, in Chapter 10 we prove that for products of random matrices in $\mathrm{GL}(2)$ the Lyapunov exponents do depend continuously on the cocycle data.

8. Each chapter ends with set of notes and a list of exercises. Some of the exercises are actually used in the proofs. They should be viewed as an invitation for the reader to take an active part in the arguments. Throughout, it is assumed that the reader is familiar with the basic ideas of Measure Theory, Differential Topology and Ergodic Theory. All that is needed can be found, for instance, in my book with Krerley Oliveira, *Fundamentos da Teoria Ergódica* [114]; a translation into English is under way.

I thank David Tranah, of Cambridge University Press, for his interest in this book and for patiently waiting for the writing to be completed. I am also grateful to Vaughn Climenhaga, and David himself, for a careful revision of the manuscript that very much helped improve the presentation.

Rio de Janeiro, March, 2014
Marcelo Viana

1

Introduction

This chapter is a kind of overture. Simplified statements of three theorems are presented that set the tone for the whole text. Much broader versions of these theorems will appear later and several other themes around them will be introduced and developed as we move on. At this initial stage we choose to focus on the following special, yet significant, setting.

Let $A_1, \ldots, A_m$ be invertible 2×2 real matrices and let $p_1, \ldots, p_m$ be positive numbers with $p_1 + \cdots + p_m = 1$. Consider

$$L^n = L_{n-1} \cdots L_1 L_0, \quad n \geq 1,$$

where the L_j are independent random variables with identical probability distributions, such that

$$\text{the probability of } \{L_j = A_i\} \text{ is equal to } p_i$$

for all $j \geq 0$ and $i = 1, \ldots, m$. In brief, our goal is to describe the (almost certain) behavior of L^n as $n \to \infty$.

1.1 Existence of Lyapunov exponents

We begin with the following seminal result of Furstenberg and Kesten [56]:

Theorem 1.1 *There exist real numbers λ_+ and λ_- such that*

$$\lim_n \frac{1}{n} \log \|L^n\| = \lambda_+ \quad \textit{and} \quad \lim_n \frac{1}{n} \log \|(L^n)^{-1}\|^{-1} = \lambda_-$$

with full probability.

The numbers λ_+ and λ_- are called *extremal Lyapunov exponents*. Clearly,

$$\lambda_+ \geq \lambda_- \tag{1.1}$$

because $\|B\| \geq \|B^{-1}\|^{-1}$ for any invertible matrix B. If B has determinant ± 1 then we even have $\|B\| \geq 1 \geq \|B^{-1}\|^{-1}$. Hence,

$$\lambda_+ \geq 0 \geq \lambda_- \tag{1.2}$$

when all matrices A_i, $1 \leq i \leq m$ have determinant ± 1.

1.2 Pinching and twisting

Next, we discuss conditions for the inequalities (1.1) and (1.2) to be strict.

Let $\mathscr{B}$ be the *monoid* generated by the matrices A_i, $i = 1, \ldots, m$; that is, the set of all products $A_{k_1} \cdots A_{k_n}$ with $1 \leq k_j \leq m$ and $n \geq 0$ (for $n = 0$ interpret the product to be the identity matrix). We say that $\mathscr{B}$ is *pinching* if for any constant $\kappa > 1$ there exists some $B \in \mathscr{B}$ such that

$$\|B\| > \kappa \|B^{-1}\|^{-1}. \tag{1.3}$$

This means that the images of the unit circle under the elements of $\mathscr{B}$ are ellipses with arbitrarily large eccentricity. See Figure 1.1.

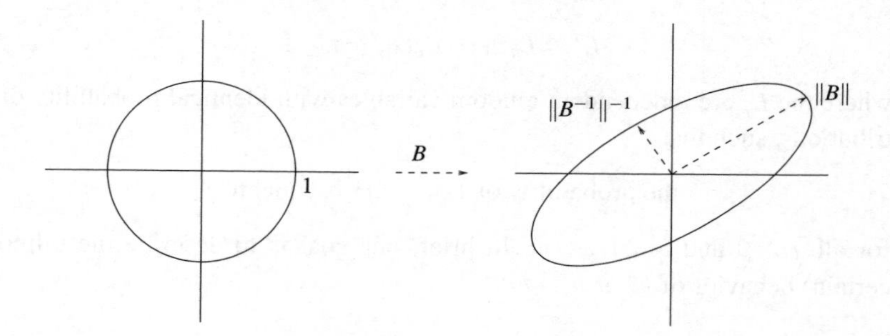

Figure 1.1 Eccentricity and pinching

We say that the monoid $\mathscr{B}$ is *twisting* if given any vector lines F, G_1, ..., $G_n \subset \mathbb{R}^2$ there exists $B \in \mathscr{B}$ such that

$$B(F) \notin \{G_1, \ldots, G_n\}. \tag{1.4}$$

The following result is a variation of a theorem of Furstenberg [54]:

Theorem 1.2 *Assume $\mathscr{B}$ is pinching and twisting. Then $\lambda_- < \lambda_+$. In particular, if $|\det A_i| = 1$ for all $1 \leq i \leq m$ then both extremal Lyapunov exponents are different from zero.*

1.3 Continuity of Lyapunov exponents

The extremal Lyapunov exponents λ_+ and λ_- may be viewed as functions of the data

$$A_1, \dots, A_m, p_1, \dots, p_m.$$

Let the matrices A_j vary in the linear group GL(2) of invertible 2×2 matrices and the probability vectors $(p_1, \dots, p_m)$ vary in the open simplex

$$\Delta^m = \{(p_1, \dots, p_m) : p_1 > 0, \dots, p_m > 0 \text{ and } p_1 + \dots + p_m = 1\}.$$

The following result is part of a theorem of Bocker and Viana [35]:

Theorem 1.3 *The extremal Lyapunov exponents $\lambda_\pm$ depend continuously on $(A_1, \dots, A_m, p_1, \dots, p_m) \in \mathrm{GL}(2)^m \times \Delta^m$ at all points.*

Example 1.4 Let $m = 2$, with

$$A_1 = \begin{pmatrix} \sigma & 0 \\ 0 & \sigma^{-1} \end{pmatrix} \quad \text{and} \quad A_2 = R_\theta A_1 R_{-\theta}, \; R_\theta = \begin{pmatrix} \cos\theta & -\sin\theta \\ \sin\theta & \cos\theta \end{pmatrix}$$

for some $\sigma > 1$ and $\theta \in \mathbb{R}$. By Theorem 1.3, the Lyapunov exponents $\lambda_\pm$ depend continuously on the parameter σ and θ. Moreover, using Theorem 1.2, we have $\lambda_+ = 0$ if and only if $p_1 = p_2 = 1/2$ and $\theta = \pi/2 + n\pi$ for some $n \in \mathbb{Z}$.

1.4 Notes

Theorem 1.1 is a special case of the theorem of Furstenberg and Kesten [56], which is valid in any dimension $d \geq 2$. The full statement and the proof will appear in Chapter 3: we will deduce this theorem from an even more general statement, the subadditive ergodic theorem of Kingman [74]. Kingman's theorem will also be used in Chapter 4 to prove the fundamental result of the theory of Lyapunov exponents, the multiplicative ergodic theorem of Oseledets [92].

It is natural to ask whether the type of asymptotic behavior prescribed by Theorem 1.1 for the norm $\|L^n\|$ and conorm $\|(L^n)^{-1}\|^{-1}$ extends to the individual *matrix coefficients* $L^n_{i,j}$. Furstenberg, Kesten [56] proved that this is so if the coefficients of the matrices A_i, $1 \leq i \leq m$ are all strictly positive. The example in Exercise 1.3 shows that this assumption cannot be removed. On the other hand, the theorem of Oseledets theorem does contain such a description for the matrix *column vectors*.

Theorem 1.2 is also the tip of a series of fundamental results, which are to be discussed in Chapters 6 through 8. The full statement and proof of Furstenberg's theorem for 2-dimensional cocycles (Furstenberg [54]) will be given in

Chapter 6. The extension to any dimension will be stated and proved in Chapter 7: it will be deduced from the invariance principle (Ledrappier [81], Bonatti, Gomez-Mont and Viana [37], Avila and Viana [16], Avila, Santamaria and Viana [13]), a general tool that has several other applications, both for linear and nonlinear systems.

In dimension larger than 2, there is a more ambitious problem: rather than asking when $\lambda_- < \lambda_+$, one wants to know when *all* the Lyapunov exponents are distinct. That will be the subject of Chapter 8, which is based on Avila and Viana [14, 15].

Furstenberg and Kifer [57] proved continuity of the Lyapunov exponents of products of random matrices, restricted to the (almost) irreducible case. A variation of their argument will be given in Section 6.2.2. The reducible case requires a delicate analysis of the random walk defined by the cocycle in projective space. That was carried out by Bocker and Viana [35], in the 2-dimensional case, using certain discretizations of projective space. At the time of writing, Avila, Eskin and Viana [12] are extending the statement of the theorem to arbitrary dimension, using a very different strategy. The proof of Theorem 1.3 that we present in Chapter 10 is based on this more recent approach.

The problem of the dependence of Lyapunov exponents on the data can be formulated in the broader context of linear cocycles that we are going to introduce in Chapter 2. We will see in Chapter 9 that, in contrast, continuity often breaks down in that generality.

1.5 Exercises

The following elementary notions are used in some of the exercises that follow. We call a 2×2 matrix *hyperbolic* if it has two distinct real eigenvalues, *parabolic* if it has a unique real eigenvalue, with a one-dimensional eigenspace, and *elliptic* if it has two distinct complex eigenvalues. Multiples of the identity belong to neither of these three classes.

Exercise 1.1 Show that, in dimension $d = 2$, if $|\det A_i| = 1$ for all $1 \leq i \leq m$ then $\lambda_+ + \lambda_- = 0$.

Exercise 1.2 Calculate the extremal Lyapunov exponents for $m = 2$ and p_1, $p_2 > 0$ with $p_1 + p_2 = 1$ and

(1) $A_1 = \begin{pmatrix} \sigma & 0 \\ 0 & \sigma^{-1} \end{pmatrix}$ and $A_2 = \begin{pmatrix} \sigma^{-1} & 0 \\ 0 & \sigma \end{pmatrix}$, where $\sigma > 1$;

(2) $A_1 = \begin{pmatrix} \sigma & 0 \\ 0 & \sigma^{-1} \end{pmatrix}$ and $A_2 = \begin{pmatrix} 0 & -1 \\ 1 & 0 \end{pmatrix}$, where $\sigma > 1$.

Exercise 1.3 (Furstenberg and Kesten [56]) Take $m = 2$ with $p_1 = p_2 = 1/2$ and

$$A_1 = \begin{pmatrix} 2 & 0 \\ 0 & 1 \end{pmatrix} \quad \text{and} \quad A_2 = \begin{pmatrix} 0 & 1 \\ 1 & 0 \end{pmatrix}.$$

Show that $\lim_n (1/n) \log |L^n_{i,j}|$ does not exist for any i, j, with full probability.

Exercise 1.4 Show that if some matrix A_i, $1 \le i \le m$ is either hyperbolic or parabolic then the monoid $\mathscr{B}$ is pinching.

Exercise 1.5 Show that the monoid $\mathscr{B}$ may be pinching even if all the matrices A_i, $1 \le i \le m$ are elliptic.

Exercise 1.6 Suppose that there exists $1 \le i \le m$ such that A_i is conjugate to an irrational rotation. Conclude that $\mathscr{B}$ is twisting.

Exercise 1.7 Suppose that there exist $1 \le i, j \le m$ such that A_i and A_j are either hyperbolic or parabolic and that they have no common eigenspace. Conclude that $\mathscr{B}$ is twisting (and pinching).

Exercise 1.8 Let A_i, $i = 1, 2$ be as in the second part of Exercise 1.2. Check that

$$\lambda_+(A_1, A_2, 1, 0) \neq \lim_{p_2 \to 0} \lambda_+(A_1, A_2, 1 - p_2, p_2).$$

Thus, the hypothesis $p_1 > 0, \ldots, p_m > 0$ cannot be removed in Theorem 1.3.

2

Linear cocycles

Linear cocycles are the basic object upon which this text is built. Here we define this concept and introduce a few examples. Special attention is given to (uniformly) hyperbolic cocycles, a class that is often used as a kind of paradigm for the behavior of more general systems.

Let $(M,\mathscr{B},\mu)$ be a probability space and $f:M\to M$ be a measure-preserving map. Let $A:M\to \mathrm{GL}(d)$ be a measurable function with values in the *linear group* $\mathrm{GL}(d)$ of invertible $d\times d$ matrices with real coefficients. Sometimes we let A take values in the *special linear group* $\mathrm{SL}(d)$ of real $d\times d$ matrices with determinant ± 1. The *linear cocycle* defined by A over f is the transformation

$$F: M\times\mathbb{R}^d \to M\times\mathbb{R}^d, \quad (x,v)\mapsto (f(x),A(x)v). \tag{2.1}$$

Observe that $F^n(x,v)=(f^n(x),A^n(x))$ for every $n\geq 1$, where

$$A^n(x)=A(f^{n-1}(x))\cdots A(f(x))A(x).$$

If f is invertible then so is F. Moreover, $F^{-n}(x,v)=(f^{-n}(x),A^{-n}(x))$ for all $n\geq 1$, where

$$A^{-n}(x)=A(f^{-n}(x))^{-1}\cdots A(f^{-1}(x))^{-1}=A^n(f^{-n}(x))^{-1}.$$

The Furstenberg–Kesten theorem (Theorem 1.1) extends to this setting, as follows: for any f-invariant probability measure μ such that $\log\|A^{\pm 1}\|\in L^1(\mu)$,

$$\lambda_+(x)=\lim_n \frac{1}{n}\log\|A^n(x)\| \quad \text{and} \quad \lambda_-(x)=\lim_n \frac{1}{n}\log\|A^n(x)^{-1}\|^{-1} \tag{2.2}$$

exist at μ-almost every point x. This fact will be proven in Chapter 3.

More generally, one may consider A to take values in the group $\mathrm{GL}(d,\mathbb{C})$ of invertible $d\times d$ matrices with complex coefficients, or the subgroup $\mathrm{SL}(d,\mathbb{C})$ of matrices with determinant in the unit circle. This gives rise to *complex linear cocycles* $M\times\mathbb{C}^d\to M\times\mathbb{C}^d$. Of course, every complex cocycle in dimension

d is also a real cocycle in dimension $2d$ and, conversely, every d-dimensional real linear cocycle defines a d-dimensional complex linear cocycle. The two theories, real and complex, are actually very similar. We focus on the real case, except where stated otherwise.

2.1 Examples

We illustrate this notion with three important classes of linear cocycles, arising from probability theory, dynamical systems, and spectral theory, respectively.

2.1.1 Products of random matrices

The situation we considered in Chapter 1 can be modeled by (a special case of) the following class of linear cocycles. Let $X = \mathrm{GL}(d)$ and $M = X^{\mathbb{Z}}$ (or $M = X^{\mathbb{N}}$) and

$$f : M \to M, \quad (\alpha_k)_k \mapsto (\alpha_{k+1})_k$$

be the shift map on X. Consider the function

$$A : M \to \mathrm{GL}(d), \quad (\alpha_k)_k \mapsto \alpha_0$$

and let $F : M \times \mathbb{R}^d \to M \times \mathbb{R}^d$ be the linear cocycle defined by A over f. Note that the kth iterate of F is given by

$$F^n\big((\alpha_k)_k, v\big) = \big((\alpha_{k+n})_k, \alpha_{n-1} \dots \alpha_1 \alpha_0 v\big).$$

Given a probability measure p in the space $\mathrm{GL}(d)$, consider the product measure $\mu = p^{\mathbb{Z}}$ (or $\mu = p^{\mathbb{N}}$), which is characterized by

$$\mu\big(\{(\alpha_k)_k : \alpha_i \in E_i, \dots, \alpha_j \in E_j\}\big) = p(E_i) \cdots p(E_j)$$

for every $i \leq j$ and any measurable sets $E_1, \dots, E_j \subset X$. It is clear that μ is invariant under the shift map.

We call a *locally constant linear cocycle* the following slightly more general construction. Let $(Y, \mathscr{Y}, q)$ be any probability space and then consider $N = Y^{\mathbb{Z}}$ endowed with the product σ-algebra $\mathscr{C} = \mathscr{Y}^{\mathbb{Z}}$ and the product measure $\nu = q^{\mathbb{Z}}$ (or $N = Y^{\mathbb{N}}$ endowed with $\mathscr{C} = \mathscr{Y}^{\mathbb{N}}$ and $\nu = q^{\mathbb{N}}$). Let $g : N \to N$ be the shift map. Moreover, let $B : N \to \mathrm{GL}(d)$ be any measurable function depending only on the zeroth coordinate; that is, of the form $B(y) = \beta(y_0)$ for some measurable function $\beta : Y \to \mathrm{GL}(d)$. Then consider the linear cocycle $G : N \times \mathbb{R}^d \to N \times \mathbb{R}^d$

defined by B over g. Note that G is semi-conjugate to a cocycle F as in the previous paragraph, with $p = \beta_* q$:

$$\begin{array}{ccc} N \times \mathbb{R}^d & \xrightarrow{G} & N \times \mathbb{R}^d \\ \Phi \downarrow & & \downarrow \Phi \\ M \times \mathbb{R}^d & \xrightarrow{F} & M \times \mathbb{R}^d \end{array}$$

with $\Phi\big((x_k)_k, v\big) = \big((\beta(x_k))_k, v\big)$. For this reason, the two cocycles are equivalent for most of our purposes.

2.1.2 Derivative cocycles

Consider a diffeomorphism $f : M \to M$ on the torus $M = \mathbb{T}^d$ of dimension $d \geq 1$. It is easy to construct smooth vector fields $X_1, \ldots, X_d$ on $\mathbb{T}^d$ such that $\{X_1(x), \ldots, X_d(x)\}$ is a basis of the tangent space $T_x M$, for every $x \in M$. One says that the torus is a *parallelizable* manifold. The *derivative cocycle* of f is

$$F : M \times \mathbb{R}^d \to M \times \mathbb{R}^d, \quad (x, v) \mapsto (f(x), A(x)v),$$

where $A(x) \in \mathrm{GL}(d)$ is the matrix, with respect to these bases, of the derivative $Df(x) : T_x M \to T_{f(x)} M$.

For more general diffeomorphisms, on non-parallelizable manifolds, the previous construction does not apply. However, one can still view the derivative map $Df : TM \to M$ as a linear cocycle, in the following more general sense. Let $\pi : \mathcal{V} \to M$ be a finite-dimensional vector bundle. This means that $\mathcal{V}$ is equipped with a family of homeomorphisms $h_\alpha : U_\alpha \times \mathbb{R}^d \to \pi^{-1}(U_\alpha)$ such that:

(i) $\{U_\alpha\}$ is an open cover of M;
(ii) $\pi \circ h_\alpha(x, v) = x$ for every $x \in U_\alpha$ and any α;
(iii) for every $x \in U_\alpha \cap U_\beta$ and any α, β, there exists a linear isomorphism $L_{\alpha,\beta}(x) : \mathbb{R}^d \to \mathbb{R}^d$ such that $h_\beta^{-1} \circ h_\alpha(x, v) = (x, L_{\alpha,\beta}(x)v)$ for every v.

The integer $d \geq 1$ is the *dimension* of the vector bundle. A *linear cocycle* on $\mathcal{V}$ over a transformation $f : M \to M$ is a measurable transformation $F : \mathcal{V} \to \mathcal{V}$ such that $\pi \circ F = f \circ \pi$ and the actions $F_x : \mathcal{V}_x \to \mathcal{V}_{f(x)}$ on the fibers are linear isomorphisms:

$$\begin{array}{ccc} \mathcal{V} & \xrightarrow{F} & \mathcal{V} \\ \pi \downarrow & & \downarrow \pi \\ M & \xrightarrow{f} & M \end{array}$$

For our purposes there is not much to gain from considering such generality. So, most of the time we will stick to the case of trivial fiber bundles; that is, to linear cocycles of the form (2.1).

2.1.3 Schrödinger cocycles

Consider $\ell^2 = \{(u_n)_{n\in\mathbb{Z}} : \sum_n |u_n|^2 < \infty\}$. The *Schrödinger operator* associated with a sequence $(V_n)_{n\in\mathbb{Z}}$ in $\mathbb{R}$, is defined by

$$H : \ell^2 \to \ell^2, \quad u = (u_n)_{n\in\mathbb{Z}} \mapsto H(u) = (u_{n+1} + u_{n-1} + V_n u_n)_{n\in\mathbb{Z}}. \tag{2.3}$$

In the most interesting models the sequence V_n is generated from a dynamical system $f : M \to M$ and a function $V : M \to \mathbb{R}$ (the so-called *potential*) through $V_n = V(f^n(x))$, for some $x \in M$. The most studied cases are:

(1) Random Schrödinger cocycles: Let $M = X^{\mathbb{Z}}$ be a shift space, $f : M \to M$ be the shift map, and $\mu = p^{\mathbb{Z}}$ be a Bernoulli measure on M. Fix $x \in M$ and then take $V_n = V(f^n(x))$, where the function $V : M \to \mathbb{R}$ is such that $V(x)$ depends only on the zeroth coordinate of $x \in M$.
(2) Quasi-periodic Schrödinger cocycles: Let μ the normalized Lebesgue measure on $M = \mathbb{T}^d$ and $f : \mathbb{T}^d \to \mathbb{T}^d$ be an irrational translation. Fix $x \in \mathbb{T}^d$ and take $V_n = V(f^n(x))$, where $V : \mathbb{T}^d \to \mathbb{R}$ is an analytic function.

A main objective is to understand the spectral theory of these operators (a theorem of Pastur [97] asserts that when the base system (f,μ) is ergodic the spectrum of H is the same for almost all choices of $x \in M$; the same is true for the absolutely continuous spectrum, the singular continuous spectrum and the pure point spectrum, by Kunz and Souillard [78]). Thus, one is led to studying the eigenvalue equation

$$H(u) = Eu, \qquad \text{for } E \in \mathbb{R}. \tag{2.4}$$

While, by definition, the eigenvectors of H are the solutions of this equation in the space ℓ^2, it is useful to consider (2.4) for any real sequence $u = (u_n)_{n\in\mathbb{Z}}$. Note that the equation may be rewritten as

$$u_{n+1} + u_{n-1} + V(f^n(x))u_n = Eu_n$$

or, still equivalently,

$$\begin{pmatrix} u_{n+1} \\ u_n \end{pmatrix} = \begin{pmatrix} E - V(f^n(x)) & -1 \\ 1 & 0 \end{pmatrix} \begin{pmatrix} u_n \\ u_{n-1} \end{pmatrix}.$$

This suggests that we consider the linear cocycle $F_E : M \times \mathbb{R}^2 \to M \times \mathbb{R}^2$ defined over $f : M \to M$ by the function

$$A : M \to \mathrm{SL}(2), \quad A(y) = \begin{pmatrix} E - V(y) & -1 \\ 1 & 0 \end{pmatrix}.$$

We have just seen that $u = (u_n)_{n\in\mathbb{Z}}$ is a solution to $H(u) = Eu$ if and only if $U = (u_n, u_{n-1})_{n\in\mathbb{Z}}$ is a trajectory of the linear cocycle F_E. The behavior of these linear cocycles provides useful information about the spectral properties of the Schrödinger operator. For example, if the Lyapunov exponents of F_E are different from zero then E cannot be an eigenvalue of $H : \ell^2 \to \ell^2$. See Damanik [48] for much more information.

2.2 Hyperbolic cocycles

We are going to define an important class of cocycles whose behavior is particularly well understood. We focus on the two-dimensional setting, but we also comment briefly on the general case.

2.2.1 Definition and properties

Let M be a compact metric space and $f : M \to M$ be a homeomorphism. We call a continuous cocycle

$$F : M \times \mathbb{R}^2 \to M \times \mathbb{R}^2, \quad (x, v) \mapsto (f(x), A(x)v)$$

hyperbolic if there are $C > 0$ and $\lambda < 1$ and, for every $x \in M$, there exist transverse lines E^s_x and E^u_x in $\mathbb{R}^2$ such that

(1) $A(x)E^s_x = E^s_{f(x)}$ and $A(x)E^u_x = E^u_{f(x)}$
(2) $\|A^n(x)v^s\| \le C\lambda^n\|v^s\|$ and $\|A^{-n}(x)v^u\| \le C\lambda^n\|v^u\|$

for every $v^s \in E^s_x$, $v^u \in E^u_x$, $x \in M$, and $n \ge 1$.

Proposition 2.1 *Let $F : M \times \mathbb{R}^2 \to M \times \mathbb{R}^2$ be the linear cocycle defined by a continuous function $A : M \to \mathrm{SL}(2)$ over a homeomorphism $f : M \to M$. Then F is hyperbolic if and only if there exist constants $c > 0$ and $\sigma > 1$ such that $\|A^n(x)\| \ge c\sigma^n$ for all $x \in M$ and $n \ge 1$.*

Proof (We are going to use (2.2), whose proof will be given in Section 3.2. Similar arguments will appear in Section 3.4, for proving the multiplicative ergodic theorem in dimension 2.)

Suppose that F is hyperbolic. Condition (2) in the definition implies that $\|A^n(x)v^u\| \geq C^{-1}\lambda^{-n}\|v^u\|$, and so

$$\|A^n(x)\| \geq C^{-1}\lambda^{-n} \quad \text{for every } x \text{ and } n.$$

This means that we may take $c = C^{-1}$ and $\sigma = \lambda^{-1}$. In the converse direction, suppose that $\|A^n(x)\| \geq c\sigma^n$ for every x and n. In particular, $\|A^n(x)\| > 1$ for every large n. Let $u_n(x)$ and $s_n(x)$ be unit vectors, respectively, most expanded and most contracted by $A^n(x)$ (Exercise 2.3):

$$\begin{aligned} &\|A^n(x)u_n(x)\| = \|A^n(x)\| \geq c\sigma^n \quad \text{and} \\ &\|A^n(x)s_n(x)\| = \|A^n(x)^{-1}\|^{-1} = \|A^n(x)\|^{-1} \leq c^{-1}\sigma^{-n}. \end{aligned} \tag{2.5}$$

Lemma 2.2 *There are C_1, $C_2 > 0$ such that*

$$|\sin\measuredangle(s_n(x), s_{n+1}(x))| \leq C_1\sigma^{-n}\|A^n(x)\|^{-1} \leq C_2\sigma^{-2n}.$$

for all $n \geq 0$ and $x \in M$.

Proof Write $\alpha_n = \measuredangle\big(s_n(x), s_{n+1}(x)\big)$. Then

$$s_n(x) = \sin\alpha_n u_{n+1}(x) + \cos\alpha_n s_{n+1}(x).$$

Then, since the images of $s_{n+1}(x)$ and $u_{n+1}(x)$ under $A^{n+1}(x)$ are orthogonal,

$$\|A^{n+1}(x)s_n(x)\| \geq \|\sin\alpha_n A^{n+1}(x)u_{n+1}(x)\| = |\sin\alpha_n|\,\|A^{n+1}(x)\|.$$

On the other hand,

$$\|A^{n+1}(x)s_n(x)\| \leq \|A(f^n(x))\|\,\|A^n(x)s_n(x)\| = \|A(f^n(x))\|\,\|A^n(x)\|^{-1}.$$

Let $C_0 > 0$ be an upper bound for the norm of A. Then, substituting (2.5),

$$|\sin\alpha_n| \leq \frac{\|A(f^n(x))\|}{\|A^{n+1}(x)\|\,\|A^n(x)\|} \leq \frac{C_0}{c\sigma^{n+1}\|A^n(x)\|} \leq \frac{C_0}{c^2\sigma^{2n+1}}.$$

This implies the conclusion of the lemma, with $c\sigma C_1 = c^2\sigma C_2 = C_0$. □

Lemma 2.2 implies that $(s_n(x))_n$ is a Cauchy sequence in projective space; that is, it can be made a Cauchy sequence by multiplying some of the vectors $s_n(x)$ by -1. Let $s(x) = \lim_n s_n(x)$. Then,

$$|\sin\measuredangle(s_n(x), s(x))| \leq C_1 \sum_{m=n}^{\infty} \sigma^{-m}\|A^m(x)\|^{-1} \leq C_2 \sum_{m=n}^{\infty} \sigma^{-2m} \tag{2.6}$$

for every $x \in M$ and every $n \geq 1$. In particular, the convergence is uniform. This also implies that $s(x)$ is a continuous function of $x \in M$.

Lemma 2.3 *$A(x)s(x)$ is collinear to $s(f(x))$ for every $x \in M$.*

Proof Let $\beta_n = \measuredangle\big(A(x)s_{n+1}(x), s_n(f(x))\big)$. Then

$$A(x)s_{n+1}(x) = \cos\beta_n s_n(f(x)) + \sin\beta_n u_n(f(x)).$$

Applying $A^n(f(x))$ to both sides,

$$\|A^{n+1}(x)s_{n+1}(x)\| \geq |\sin\beta_n|\,\|A^n(f(x))u_n(f(x))\| - \|A^n(f(x))s_n(f(x))\|.$$

Substituting (2.5), we get that $c^{-1}\sigma^{-n-1} \geq c\sigma^n|\sin\beta_n| - c^{-1}\sigma^{-n}$, and so

$$|\sin\beta_n| \leq 2c^{-2}\sigma^{-2n}.$$

So $|\sin\beta_n| \to 0$ as $n \to \infty$, and this implies the claim in the lemma. □

Lemma 2.4 *For any $\sigma_0 < \sigma$ there exists $n_0 \geq 1$ such that $\|A^n(x)s(x)\| \leq \sigma_0^{-n}$ for every $x \in M$ and $n \geq n_0$.*

Proof First, take $x \in M$ to be such that $\lim_n n^{-1}\log\|A^n(x)\|$ exists. Observe that, according to (2.2), this is the case for μ-almost every point x and every f-invariant probability measure μ. We claim that

$$\limsup_n \frac{1}{n}\log\|A^n(x)s(x)\| \leq -\log\sigma. \tag{2.7}$$

For proving this, let $\gamma_n = \measuredangle\big(s(x), s_n(x)\big)$. Then $s(x) = \cos\gamma_n s_n(x) + \sin\gamma_n u_n(x)$ and so, using (2.5) and (2.6),

$$\begin{aligned}\|A^n(x)s(x)\| &\leq |\cos\gamma_n|\|A^n(x)s_n(x)\| + |\sin\gamma_n|\|A^n(x)u_n(x)\|\\ &\leq \|A^n\|^{-1} + C_1\sum_{m=n}^{\infty}\sigma^{-j}\|A^j\|^{-1}\|A^n\|.\end{aligned}$$

The assumption implies that for any $\varepsilon > 0$ there exists $n_\varepsilon \geq 1$ such that

$$e^{-n\varepsilon} \leq \|A^m\|^{-1}\|A^n\| \leq e^{n\varepsilon} \quad \text{for every } m \geq n \geq n_\varepsilon.$$

Using (2.5) once more, it follows that

$$\|A^n(x)s(x)\| \leq c^{-1}\sigma^{-n} + C_1e^{n\varepsilon}\sum_{j=n}^{\infty}\sigma^{-j} \leq C_1'e^{n\varepsilon}\sigma^{-n} \quad \text{for every } n \geq n_\varepsilon,$$

where the constant C_1' depends only on c, C_1, and σ. This implies our claim. Thus, we have shown that (2.7) holds for μ-almost every x and any f-invariant probability measure μ.

Now, suppose that the conclusion of the lemma is false. Then there exists $\sigma_0 < \sigma$ and, for every $k \geq 1$, there exists $n_k \geq k$ and $x_k \in M$ such that

$$\|A^{n_k}(x_k)s(x_k)\| > \sigma_0^{-n_k}.$$

Define $\mu_k = n_k^{-1}\sum_{j=0}^{n_k-1}\delta_{f^j(x_k)}$ and $\phi(x) = \log\|A(x)s(x)\|$. The previous relation may be written, equivalently, as $\int \phi \, d\mu_k > -\log \sigma_0$. Since the space of probability measures on M is weak*-compact, up to restricting to a subsequence if necessary we may assume that the sequence $(\mu_k)_k$ converges in the weak* topology to some probability measure μ on M. Since the function is continuous, it follows that

$$\int \phi \, d\mu \geq -\log \sigma_0.$$

Clearly, $f_*\mu_k = \mu_k + n_k^{-1}\big(\delta_{f^{n_k}(x_k)} - \delta_{x_k}\big)$. Making $k \to \infty$ we find that $f_*\mu = \mu$; that is, μ is invariant under f. Let $\tilde{\phi}$ be the Birkhoff time average of ϕ:

$$\tilde{\phi}(x) = \lim_n \frac{1}{n} \sum_{j=0}^{n-1} \phi(f^j(x)) = \lim_n \frac{1}{n} \log \|A^n(x)s(x)\|.$$

On the one hand, by the ergodic theorem, $\int \tilde{\phi} \, d\mu = \int \phi \, d\mu \geq -\log \sigma_0$. On the other hand, (2.7) gives that $\tilde{\phi}(x) \leq -\log \sigma$ for μ-almost every x. These two inequalities are incompatible. This contradiction proves the lemma. □

Let $\sigma_0 \in (1, \sigma)$ be fixed. By Lemma 2.4, there exists $C_3 > 0$ such that

$$\|A^n(x)s(x)\| \leq C_3 \sigma_0^{-n} \quad \text{for every } x \in M \text{ and every } n \geq 1. \tag{2.8}$$

Analogously, considering backward iterates instead, one constructs a unit vector $u(x)$ such that $A^{-1}(x)u(x)$ is collinear to $u(x)$ for every $x \in M$ and

$$\|A^{-n}(x)u(x)\| \leq C_3 \sigma_0^{-n} \quad \text{for every } x \in M \text{ and every } n \geq 1. \tag{2.9}$$

Incidentally, this is the first time in the proof that we have used the assumption that the cocycle is invertible.

Lemma 2.5 *The vectors $s(x)$ and $u(x)$ are transverse for every $x \in M$.*

Proof From (2.8) we get that $\|A^n(f^{-n}(x))s(f^{-n}(x))\| \leq C_3 \sigma^{-n}$ for every x and n. This may be rewritten as

$$\|A^{-n}(x)s(x)\| \geq C_3^{-1} \sigma_0^n,$$

in view of Lemma 2.3. Moreover, using (2.9),

$$\|A^{-n}(x)u(x)\| \leq C_3 \sigma_0^{-n},$$

for every x and n. Then $\|A^{-n}(x)s(x)\|$ is larger than $\|A^{-n}(x)u(x)\|$ for every large n. In particular, $s(x)$ and $u(x)$ cannot be collinear, as claimed. □

Define E_x^s and E_x^u to be the lines generated by $s(x)$ and $u(x)$, respectively. Lemmas 2.2–2.5 yield all the properties in the conclusion of Proposition 2.1. □

2.2.2 Stability and continuity

Let $C^0(M, \mathrm{SL}(2))$ denote the space of continuous functions $M \to \mathrm{SL}(2)$, endowed with the distance

$$d(A,B) = \sup_{x \in M} \|A(x) - B(x)\|.$$

We are going to deduce from Proposition 2.1 that hyperbolicity corresponds to an open subset of this space. Moreover, the invariant decomposition varies continuously on this subset.

Proposition 2.6 *Let $f : M \to M$ be fixed. Suppose that the linear cocycle F defined by $A : M \to \mathrm{SL}(2)$ over f is hyperbolic, and let $\mathbb{R}^2 = E^s_x \oplus E^u_x$ be the corresponding invariant decomposition.*

There exists $\delta > 0$ such that the cocycle defined over f by any continuous function $B : M \to \mathrm{SL}(2)$ with $d(A,B) < \delta$ is also hyperbolic.

Moreover, given $\varepsilon > 0$, the B-invariant decomposition $\mathbb{R}^2 = E^s_{B,x} \oplus E^u_{B,x}$, $x \in M$ satisfies $|\sin\measuredangle(E^s_x, E^s_{B,x})| < \varepsilon$ and $|\sin\measuredangle(E^u_x, E^u_{B,x})| < \varepsilon$ for all $x \in M$, if δ is small enough.

Proof For each $x \in M$ and $\gamma > 0$, let $C^u(x,\gamma)$ be the set of vectors $v \in \mathbb{R}^d$ whose coordinates (v^s, v^u) in the decomposition $\mathbb{R}^d = E^s_x \oplus E^u_x$ satisfy $\|v^s\| \le \gamma\|v^u\|$ (this is a particular case of a *cone*, of which we will hear more in Section 4.4.2). Let $C > 0$ and $\lambda < 1$ be constants as in the definition of hyperbolicity. Fix $k \ge 1$ large enough so that $C\lambda^k \le 1/3$. Then, for any $v \in C^u(x,1)$ and $x \in M$,

$$\|A^k(x)v^u\| \ge 3\|v^u\| \quad \text{and} \quad \|A^k(x)v^s\| \le \frac{1}{3}\|v^s\| \le \frac{1}{3}\|v^u\| \le \frac{1}{9}\|A^k(x)v^u\|,$$

and so $A^k(x)v \in C^u(x,1/9)$. Then, for any $B : M \to SL(2)$ such that $d(A,B)$ is sufficiently small, for every $v \in C^u(x,1)$, and every $x \in M$,

$$\|\big(B^k(x)v\big)^u\| \ge 2\|v^u\| \quad \text{and} \quad B^k(x)v \in C^u(x,1).$$

So, by induction, $\|\big(B^{kn}(x)v\big)^u\| \ge 2^n\|v^u\|$ for every $v \in C^u(x,1)$, every $x \in M$, and every $n \ge 1$. This implies that $\|B^n(x)\| \ge c\sigma^n$ for every $x \in M$ and $n \ge 1$, where $\sigma = 2^{1/k}$ and $c > 0$ are independent of B, x, and n. By Proposition 2.1, it follows that the cocycle defined by B is hyperbolic, as claimed.

To prove the other claim in the proposition, let $\varepsilon > 0$ be fixed. Let $s(x)$ be a unit vector in the direction of $E^s_{B,x}$ and, for each $n \ge 1$, let $s_{B,n}(x)$ be the unit vector most contracted by $B^n(x)$. By (2.6), there is a constant $C > 0$ independent of B, x, and n, such that

$$|\sin\measuredangle(s_{B,n}(x), s_B(x))| \le C\sigma^{-2n}.$$

Fix $n \geq 1$, large enough so that $C\sigma^{-2n} < \varepsilon/4$ and then assume $d(A,B)$ is small enough so that $|\sin\measuredangle(s_{A,n}(x), s_{B,n}(x))| < \varepsilon/4$ for every $x \in M$ and $n \geq 1$. Then

$$|\sin\measuredangle(s_A(x), s_B(x))| \leq |\sin\measuredangle(s_{A,n}(x), s_{B,n}(x))| + 2C\sigma^{-2n} < \varepsilon.$$

This proves that $|\sin\measuredangle(E^s_x, E^s_{B,x})| < \varepsilon$ for every $x \in M$. The corresponding statement for E^u is analogous. □

Here is another interesting consequence of Proposition 2.1.

Proposition 2.7 *Suppose that $A : M \to \mathrm{SL}(2)$ is such that $A(x)$ has positive entries for every $x \in M$. Then the linear cocycle F defined by A over any homeomorphism $f : M \to M$ is hyperbolic.*

Proof By hypothesis,

$$A(x) = \begin{pmatrix} a_x & b_x \\ c_x & d_x \end{pmatrix}$$

with a_x, b_x, c_x, $d_x > 0$ and $a_x d_x - b_x c_x = 1$. Since M is compact and A is continuous, there exists a positive lower bound $\delta > 0$ for a_x, b_x, c_x, and d_x over all $x \in M$. It is clear that the cocycle preserves the subset of vectors with positive entries: if $v = (v_0, w_0)$ is such that $v_0 > 0$ and $w_0 > 0$ then $A(x)v = (v_1, w_1)$ with $v_1 > 0$ and $w_1 > 0$. Moreover,

$$v_1 w_1 > (1 + 2b_x c_x) v_0 w_0 \geq (1 + 2\delta^2) v_0 w_0.$$

Then, by induction, the iterates $(v_n, w_n) = A^n(x)v$ have $v_n > 0$, $w_n > 0$ and $v_n w_n > (1 + 2\delta^2)^n v_0 w_0$. This implies that the norm of $A^n(x)$ grows exponentially fast: $\|A^n(x)\| \geq c\sigma^n$ for some $c > 0$ and $\sigma = (1 + 2\delta^2)^{1/2}$. Now the claim follows directly from Proposition 2.1. □

The notion of hyperbolicity extends naturally to linear cocycles in any dimension and one may even consider linear cocycles on general vector bundles, as mentioned in Section 2.1.2. Namely, let $f : M \to M$ be a homeomorphism on a compact space M. A continuous linear cocycle $F : \mathcal{V} \to \mathcal{V}$ over f is *hyperbolic* if there are constants $C > 0$ and $\lambda < 1$ and for every $x \in M$ there exists a direct sum decomposition $\mathcal{V}_x = E^s_x \oplus E^u_x$ of the fiber over x satisfying

(1) $F_x(E^s_x) = E^s_{f(x)}$ and $F_x(E^u_x) = E^u_{f(x)}$
(2) $\|F^n_x(v^s)\| \leq C\lambda^n \|v^s\|$ and $\|F^{-n}_x(v^u)\| \leq C\lambda^n \|v^u\|$ for every $v^s \in E^s_x$, $v^u \in E^u_x$, and $n \geq 1$.

The conclusion of Proposition 2.6 remains valid in this generality, as we are going to explain. First of all, there is a natural topology in the space of linear cocycles over $f : M \to M$, associated with the following norm. Fix any finite set of trivializing coordinates $h_\alpha : U_\alpha \times \mathbb{R}^d \to \pi^{-1}(U_\alpha)$ for the vector bundle $\mathscr{V}$. The local expressions

$$h_\beta^{-1} \circ F \circ h_\alpha : \left(U_\alpha \cap f^{-1}(U_\beta)\right) \times \mathbb{R}^d \to U_\beta \times \mathbb{R}^d$$

of F are maps of the form $(x,v) \mapsto (f(x), F_{\alpha,\beta}(x)v)$. Define

$$\|F\| = \max_{\alpha,\beta} \sup \left\{ \|F_{\alpha,\beta}(x)\| : x \in U_\alpha \cap f^{-1}(U_\beta) \right\}.$$

The topology associated with this norm does not depend on the choice of the trivializing atlas.

Then the set of hyperbolic cocycles is an open subset of the space of all linear cocycles over $f : M \to M$. Moreover, the invariant sub-bundles E^s and E^u vary continuously with the linear cocycle inside that open set, in the following sense: restricted to the each trivializing domain U_α, we may view $x \mapsto E^*_x$, $* \in \{s,u\}$ as a map from U_α to the Grassmannian $\mathrm{Gr}(d)$; if F is a hyperbolic linear cocycle and G is a nearby linear cocycle then the maps $U_\alpha \to \mathrm{Gr}(d)$ associated with G are uniformly close to those of F, for each α. The proof of these facts is part of Exercise 2.6 (see also Shub [107]).

By the *Grassmannian* $\mathrm{Gr}(d)$ we mean the (disjoint) union of the *Grassmannian manifolds* $\mathrm{Gr}(l,d)$, $0 \le l \le d$, whose elements are the l-dimensional vector subspaces of $\mathbb{R}^d$.

2.2.3 Obstructions to hyperbolicity

We describe a few mechanisms that exclude the presence of hyperbolic cocycles in certain situations.

Example 2.8 Let M be a compact, connected metric space, $f : M \to M$ be a continuous transformation, and $A : M \to \mathrm{SL}(d)$ be a continuous function such that for every $1 \le i \le d-1$ there exists a periodic point $p_i \in M$ of f, with period $\kappa_i \ge 1$ such that the eigenvalues $\{\beta^i_j : 1 \le j \le d\}$ of each $A^{\kappa_i}(p_i)$ satisfy

$$|\beta^i_1| \ge \cdots \ge |\beta^i_{i-1}| > |\beta^i_i| = |\beta^{i+1}_i| > |\beta^i_{i+2}| \ge \cdots \ge |\beta^i_d| \tag{2.10}$$

and β^i_i, β^i_{i+1} are complex conjugate (not real). Such an A may be found, for instance, starting with a constant cocycle and deforming it on disjoint neighborhoods of the periodic orbits. Property (2.10) remains valid for every $B : M \to \mathrm{SL}(d)$ in a C^0 neighborhood $\mathscr{U}$ of A. Consider any $B \in \mathscr{U}$ and suppose the associated cocycle is hyperbolic, with hyperbolic decomposition $E^s_x \oplus E^u_x$ at each $x \in M$. Since the hyperbolic decomposition is continuous (Exercises 2.1 and 2.6), and M is connected, the dimensions of E^s_x and E^u_x are constant. Let $\dim E_x = l$. Then, for any point $p \in M$ with $f^\kappa(p) = p$, the l largest (in norm) eigenvalues of $B^\kappa(p)$ are strictly larger than the other $d-l$ eigenvalues. This is incompatible with (2.10) when $p = p_l$. This contradiction shows that no cocycle in $\mathcal{U}$ can be hyperbolic.

Example 2.9 (Michael Herman) Let $S^1 = \mathbb{R}/\mathbb{Z}$ and $f : S^1 \to S^1$ be a continuous transformation, with topological degree $\deg(f) \in \mathbb{Z}$. Let $A : S^1 \to \mathrm{SL}(2)$ be of the form $A(x) = A_0 R_{2\pi\alpha(x)}$ where $A_0 \in \mathrm{SL}(d)$, $\alpha : S^1 \to S^1$ is a continuous function with topological degree $\deg(\alpha) \in \mathbb{Z}$, and R_θ denotes the rotation of angle θ. Let $\mathcal{U}$ be isotopy class of A in the space of maps from S^1 to $\mathrm{SL}(2)$. This is a C^0 neighborhood of A. We claim that *if* $2\deg(\alpha)$ *is not a multiple of* $\deg(f) - 1$ *then the linear cocycle associated with every* $B \in \mathcal{U}$ *is not hyperbolic*. Indeed, let $B \in \mathcal{U}$ and suppose the associated cocycle is hyperbolic. Let $E^s_x \oplus E^u_x$ be the hyperbolic decomposition. Let us view $E^s : x \mapsto E^s_x$ as a continuous map from S^1 to the projectivization $\mathbb{PR}^2$ of the real plane. The graph $\{(x, E^s_x) : x \in S^1\}$ represents some element (η, ζ) of the fundamental group $\pi_1(S^1 \times \mathbb{PR}^2) = \mathbb{Z} \oplus \mathbb{Z}$. Since B is isotopic to A, the image of $\mathrm{graph}(E^s)$ under the cocycle must represent

$$(\eta \deg(f), \zeta + 2\deg(\alpha)) \in \pi_1(S^1 \times \mathbb{PR}^2)$$

(the factor 2 comes from the fact that S^1 is the 2-fold covering of $\mathbb{PR}^2$). This must be collinear to (η, ζ), because E^s is invariant under the cocycle. In other words, we must have $\zeta + 2\deg(\alpha) = \deg(f)\zeta$. Then $\deg(f) - 1$ must divide $2\deg(\alpha)$. This proves our claim.

Example 2.10 A diffeomorphism $f : M \to M$ on a compact manifold is an *Anosov diffeomorphism* if the derivative cocycle $F = Df$ is hyperbolic. The existence of Anosov diffeomorphisms imposes strong restrictions on the manifold M. For one thing, the Euler characteristic must be zero. In dimension 2 the Klein bottle may also be excluded, so that the only surface that admits Anosov diffeomorphisms is the torus $\mathbb{T}^2$.

Anosov diffeomorphisms can also be constructed in the high-dimensional tori, as follows. Let A be any $d \times d$ a matrix with integer coefficients and determinant 1. Then A defines a diffeomorphism of $\mathbb{T}^d = \mathbb{R}^d/\mathbb{Z}^d$ and, assuming that A has no eigenvalues in the unit circle, this diffeomorphism is Anosov. Every Anosov diffeomorphism on a torus is topologically conjugate to such a hyperbolic automorphism.

Anosov diffeomorphisms with a similar algebraic flavor may be constructed on the more general class of infranilmanifolds, which are suitable quotients of Lie groups. Again, all Anosov diffeomorphisms on infranilmanifolds are topologically conjugate to a hyperbolic automorphism of the manifold. Moreover, all known examples of Anosov diffeomorphisms are defined on infranilmanifolds. See Newhouse [91], Franks [52], Manning [90].

2.3 Notes

The theory of products of random matrices was effectively initiated by Furstenberg and Kesten [56] and Furstenberg [54]. There is now a vast literature on the subject, some of which will be discussed later. In addition to the references we provide along the way, let us mention the collections [7, 44, 51] and the book [46].

The time-independent Schrödinger equation describes the so-called *orbitals*, or *stationary waves*, in quantum mechanics. It takes the form $H\Psi = E\psi$ where Ψ is the wave function, E is the energy of the quantum state Ψ and H is the *Hamiltonian operator*, an hermitian operator that characterizes the total energy of any wave function. In the case of a single particle moving in an electric field the Hamiltonian operator is given by

$$H\Psi(x) = \Big[-\frac{\hbar^2}{2m}\Delta + V(x) \Big] \Psi(x)$$

where $\hbar$ is the reduced Planck constant, m is the mass, V is the potential energy and Δ is the Laplacian operator. Our discussion in Section 2.1.3 concerns the case when the space variable x is one-dimensional and discrete. Then, the Laplacian is simply given by $\Delta\Psi(n) = \Psi(n+1) - 2\Psi(n) + \Psi(n-1)$ and this readily leads to the form of the Schrödinger operator in (2.3).

Much of the mathematical study of Schrödinger operators is motivated by an observation made in 1958 by the American physicist Philip Anderson. He argued in [1] that, while ideal crystals are always conductors, the presence of impurities should cause the crystal to loose all its conductivity properties and, thus, become an insulator: the electrons are trapped due to the crystal lattice disorder (this discovery earned Anderson the Nobel Prize for Physics in 1977).

Mathematically, such disordered systems are modeled by suitable (random) Schrödinger operators $H : \ell^2(\mathbb{Z}^d) \to \ell^2(\mathbb{Z}^d)$ and the Anderson localization phenomenon corresponds to the operator H having only a pure point spectrum. In other words, localization means that the space $\ell^2(\mathbb{Z}^d)$ admits a Hilbert basis formed by eigenvectors of H. See Damanik [47] and Jitomirskaya and Marx [66] for surveys on this and related problems. In the text, we restricted ourselves to hinting (in the case $d = 1$) that the Lyapunov exponents of the associated Schrödinger cocycles have a say in the spectral theory of these operators. Damanik [48] contains a lot more information on this topic.

The observations in Section 2.2 will be useful in Chapter 9. The notion of (uniform) hyperbolicity is due to Smale; see [109] and references therein. He had in mind diffeomorphisms and smooth flows (derivative cocycles) and the purpose was twofold: to prove that hyperbolicity characterizes the structural stability of the dynamics (the Stability Conjectures of Palis and Smale [96]), and to conclude that most smooth systems are structurally stable (which turned out not to be true).

Of course, hyperbolicity was implicit in some previous works, for instance in the proof by E. Hopf [65] that the geodesic flow on any surface with negative curvature is ergodic. Anosov [2] introduced the class of globally hyperbolic diffeomorphisms and flows to extend this result to arbitrary dimension: he observed that the geodesic flow on any compact manifold with negative (sectional) curvature is globally hyperbolic and he proved that volume-preserving globally hyperbolic C^2 diffeomorphisms and flows are ergodic.

Especially in the context of diffeomorphisms and smooth flows, the concept of (uniform) hyperbolicity has been broadened to that of *non-uniform hyperbolicity*, which refers to systems whose Lyapunov exponents are non-zero at almost every point, relative to some distinguished invariant measure. This was initiated by Pesin [99, 100] with major contributions also by Katok [68], Mañé [86], Ledrappier [79], Ledrappier and Young [83, 84] and Barreira, Pesin and Schmeling [19] among others. See also the book of Barreira and Pesin [18] and references therein. For most of these results the system is assumed to be $C^{1+\text{Hölder}}$.

2.4 Exercises

Exercise 2.1 Show that E^s_x and E^u_x are unique when they exist. Moreover, show they depend continuously on the point $x \in M$.

Exercise 2.2 Prove that if the Schrödinger cocycle F_E is hyperbolic then the spectral equation $H(u) = Eu$ has no solution in ℓ^2.

Exercise 2.3 Show that given $B \in \mathrm{SL}(2)$ such that $\|B\| \neq 1$, there exist unit vectors s and u such that $\|B(u)\| = \|B\|$ and $\|B(s)\| = \|B^{-1}\|^{-1} = \|B\|^{-1}$. These vectors are unique, up to multiplication by -1, they are orthogonal, and their images $B(s)$ and $B(u)$ are also orthogonal.

Exercise 2.4 Check that $\|B|r_1\|/\|B|r_2\| \leq \|B\|^2$ for any lines $r_1, r_2 \subset \mathbb{R}^2$ and $B \in \mathrm{SL}(2)$.

Exercise 2.5 Prove that a linear cocycle F is hyperbolic if and only if any iterate F^k, $k \neq 0$ is hyperbolic.

Exercise 2.6 Extend Exercise 2.1, Proposition 2.6, and Exercise 2.5 to linear cocycles on vector bundles of arbitrary dimension.

3
Extremal Lyapunov exponents

As announced already in (2.2), the theorem of Furstenberg and Kesten [56] states that, under a suitable integrability assumption,

$$\text{the norm } \|A^n(x)\| \quad \text{and} \quad \text{the conorm } \|A^n(x)^{-1}\|^{-1}$$

of any linear cocycle have well-defined exponential rates as time $n \to \infty$, for almost every point x. The precise statement, which contains Theorem 2.2, is given in Section 3.2. Indeed, we obtain the result as a straightforward consequence of the so-called subadditive ergodic theorem of Kingman [74], which we state and prove in Section 3.1.

We take the occasion, in Section 3.3, to illustrate Herman's [62] subharmonic trick for bounding the largest Lyapunov exponents from below.

The multiplicative ergodic theorem of Oseledets [92] improves the Furstenberg–Kesten theorem in that it provides exponential rates for the iterates

$$\|A^n(x)v\|$$

of all vectors, rather than just for the norm and conorm of the matrices. The full statements will appear in Chapter 4. Here (Section 3.4), we treat the special case of cocycles in dimension 2, which is much simpler and contains a few of the elements of the general case.

3.1 Subadditive ergodic theorem

Let $(M, \mathscr{B}, \mu)$ be a probability space and $f : M \to M$ be a measure-preserving transformation. The *positive part* and the *negative part* of a measurable function $\varphi : M \to [-\infty, +\infty]$ are the non-negative functions defined by, respectively,

$$\varphi^+(x) = \max\{0, \varphi(x)\} \quad \text{and} \quad \varphi^-(x) = \max\{0, -\varphi(x)\}.$$

A measurable function φ is called *(essentially) invariant* if $\varphi(f(x)) = \varphi(x)$ for μ-almost all $x \in M$. By definition, a measurable subset of M is invariant if its characteristic function is invariant. A sequence $\varphi_n : M \to [-\infty, +\infty)$, $n \geq 1$, of measurable functions is *subadditive*, relative to f, if

$$\varphi_{m+n} \leq \varphi_m + \varphi_n \circ f^m \quad \text{for all } m, n \geq 1.$$

Example 3.1 Given any measurable function $\psi : M \to \mathbb{R}$, consider its *orbital sum* $\varphi_n = \sum_{j=0}^{n-1} \psi \circ f^j$. Then $\varphi_{m+n} = \varphi_m + \varphi_n \circ f^m$ for every m and n and, in particular, $(\varphi_n)_n$ is a subadditive sequence, and even an *additive* sequence, since the equality always holds. It is clear that, conversely, every additive sequence is the orbital sum of its first term.

Example 3.2 Given any measurable function $A : M \to \mathrm{GL}(d)$, consider the sequence $\varphi_n(x) = \log \|A^n(x)\|$, where $A^n(x) = A(f^{n-1}(x)) \cdots A(f(x))A(x)$. As $\|B_1 B_2\| \leq \|B_1\| \, \|B_2\|$ for every $B_1, B_2 \in \mathrm{GL}(d)$, the sequence $(\varphi_n)_n$ is subadditive.

Theorem 3.3 (Kingman) *Let $\varphi_n : M \to [-\infty, +\infty)$, $n \geq 1$ be a subadditive sequence of measurable functions such that $\varphi_1^+ \in L^1(\mu)$. Then $(\varphi_n/n)_n$ converges μ-almost everywhere to some invariant function $\varphi : M \to [-\infty, +\infty)$. Moreover, the positive part φ^+ is integrable and*

$$\int \varphi \, d\mu = \lim_n \frac{1}{n} \int \varphi_n \, d\mu = \inf_n \frac{1}{n} \int \varphi_n \, d\mu \in [-\infty, +\infty).$$

The proof of this theorem will be presented later. An interesting feature is that it does *not* use the ergodic theorem of Birkhoff. Thus, the latter can be obtained as a consequence (Corollary 3.10).

3.1.1 Preparing the proof

A sequence $(a_n)_n$ in $[-\infty, +\infty)$ is said to be *subadditive* if $a_{m+n} \leq a_m + a_n$ holds for any $m, n \geq 1$.

Lemma 3.4 *If $(a_n)_n$ is a subadditive sequence then*

$$\lim_n \frac{a_n}{n} = \inf_n \frac{a_n}{n} \in [-\infty, \infty). \tag{3.1}$$

Proof If $a_m = -\infty$ for some m then, by subadditivity, $a_n = -\infty$ for all $n > m$. Then both sides of (3.1) are equal to $-\infty$, and so the lemma holds in this case. From now, assume that $a_n \in \mathbb{R}$ for all n.

Let $L = \inf_n (a_n/n) \in [-\infty, +\infty)$ and let L' be any real number bigger than L. Then we can find $k \geq 1$ such that $a_k/k < L'$. For $n > k$, we may write $n = kp + q$, where p and q are integer numbers such that $p \geq 1$ and $1 \leq q \leq k$. Then, by subadditivity,

$$a_n \leq a_{kp} + a_q \leq pa_k + a_q \leq pa_k + \alpha,$$

where $\alpha = \max\{a_i : 1 \leq i \leq k\}$. Then,

$$\frac{a_n}{n} \leq \frac{pk}{n}\frac{a_k}{k} + \frac{\alpha}{n}.$$

Observe that pk/n converges to 1 and α/n converges to zero when $n \to \infty$. Therefore, since $a_k/k < L'$, we have

$$L \leq \frac{a_n}{n} < L'$$

for any large enough n. Making $L' \to L$, we conclude that

$$\lim_n \frac{a_n}{n} = L = \inf_n \frac{a_n}{n}.$$

This completes the argument. □

Now let $(\varphi_n)_n$ be as in Theorem 3.3. By subadditivity,

$$\varphi_n \leq \varphi_1 + \varphi_1 \circ f + \cdots + \varphi_1 \circ f^{n-1}.$$

This inequality remains true if we replace φ_n and φ_1 by φ_n^+ and φ_1^+, respectively. So, the hypothesis that $\varphi_1^+ \in L^1(\mu)$ implies that $\varphi_n^+ \in L^1(\mu)$ for any n. On the other hand, the hypothesis that $(\varphi_n)_n$ is subadditive implies that

$$a_n = \int \varphi_n \, d\mu, \quad n \geq 1,$$

is a subadditive sequence in $[-\infty, +\infty)$. Then, by Lemma 3.4,

$$\lim_n \frac{a_n}{n} = \inf_n \frac{a_n}{n} = L \in [-\infty, \infty)$$

exists. Define $\varphi_- : M \to [-\infty, \infty]$ and $\varphi_+ : M \to [-\infty, \infty]$ by

$$\varphi_-(x) = \liminf_n \frac{\varphi_n}{n}(x) \quad \text{and} \quad \varphi_+(x) = \limsup_n \frac{\varphi_n}{n}(x).$$

It is clear that $\varphi_-(x) \leq \varphi_+(x)$ for every $x \in M$. We are going to prove that

$$\int \varphi_- \, d\mu \geq L \geq \int \varphi_+ \, d\mu, \tag{3.2}$$

provided that every function φ_n is bounded from $-\infty$. Consequently, the two functions φ_- and φ_+ coincide at μ-almost every point and their integrals are equal to L. This yields the theorem in this case, with $\varphi = \varphi_- = \varphi_+$ (the fact that φ is an invariant function is proved in Exercise 3.2). At the end, we use a truncation trick to remove the boundedness condition and, hence, prove the theorem in complete generality.

3.1.2 Fundamental lemma

In this section we assume that $\varphi_- > -\infty$ at every point. Fix $\varepsilon > 0$ and define, for each $k \in \mathbb{N}$,

$$E_k = \{x \in M : \varphi_j(x) \le j(\varphi_-(x) + \varepsilon) \text{ for some } j \in \{1, \dots, k\}\}.$$

It is clear that $E_k \subset E_{k+1}$ for any k. Moreover, the definition of $\varphi_-(x)$ implies that $M = \bigcup_k E_k$. Define also,

$$\psi_k(x) = \begin{cases} \varphi_-(x) + \varepsilon & \text{if } x \in E_k \\ \varphi_1(x) & \text{if } x \in E_k^c. \end{cases}$$

The definition of E_k implies that $\varphi_1(x) > \varphi_-(x) + \varepsilon$ for every $x \in E_k^c$. Thus, $\psi_k(x)$ decreases to $\varphi_-(x) + \varepsilon$ as $k \to \infty$, for every $x \in M$. In particular, by the monotonic convergence theorem,

$$\int \psi_k \, d\mu \to \int (\varphi_- + \varepsilon) \, d\mu \quad \text{as } k \to \infty.$$

The crucial step in the proof of the theorem is the following estimate:

Lemma 3.5 *For any $n > k \ge 1$ and μ-almost every point $x \in M$,*

$$\varphi_n(x) \le \sum_{i=0}^{n-k-1} \psi_k(f^i(x)) + \sum_{i=n-k}^{n-1} \max\{\psi_k, \varphi_1\}(f^i(x)).$$

This means that the sequence $(\varphi_n)_n$ is bounded from above by an orbital sum of ψ_k and an "error term". Since orbital sums are additive sequences (recall Example 3.1), this pretty much reduces our subadditive setting to the much easier additive case. Recall that ψ_k converges to $\varphi_- + \varepsilon$ when k goes to infinity. The last sum on the right-hand side of the inequality (the "error term") is negligible because it is the sum of a fixed number k of integrable functions, and n may be taken to be much larger than k.

Proof Take $x \in M$ such that $\varphi_-(x) = \varphi_-(f^j(x))$ for any $j \ge 1$ (this is the case for μ-almost every point; see Exercise 3.2). Consider the sequence, possibly finite, of integer numbers

$$m_0 \le n_1 < m_1 \le n_2 < m_2 < \cdots \tag{3.3}$$

defined inductively in the following way (see also Figure 3.1).

Take $m_0 = 0$. Given $j \ge 1$, let n_j be the smallest integer greater or equal than m_{j-1} that satisfies $f^{n_j}(x) \in E_k$ (assuming it exists). Then, by definition of E_k, there exists m_j such that $1 \le m_j - n_j \le k$ and

$$\varphi_{m_j - n_j}(f^{n_j}(x)) \le (m_j - n_j)(\varphi_-(f^{n_j}(x)) + \varepsilon). \tag{3.4}$$

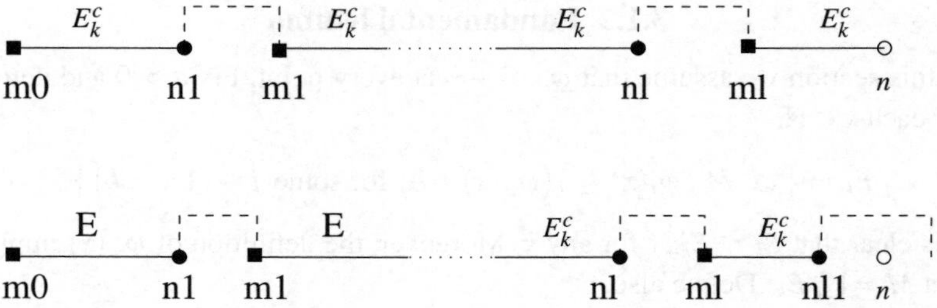

E_k^c E_k^c E_k^c E_k^c

m0 n1 m1 nl ml n

E E E_k^c E_k^c

m0 n1 m1 nl ml nl1 n

Figure 3.1 Splitting the trajectory of a point

This completes the definition of the sequence (3.3). Given any $n \geq k$, let $l \geq 0$ be largest such that $m_l \leq n$. By subadditivity,

$$\varphi_{n_j - m_{j-1}}(f^{m_{j-1}}(x)) \leq \sum_{i=m_{j-1}}^{n_j - 1} \varphi_1(f^i(x))$$

for any $j = 1, \ldots, l$ such that $m_{j-1} \neq n_j$, and similarly for $\varphi_{n-m_l}(f^{m_l}(x))$. Thus,

$$\varphi_n(x) \leq \sum_{i \in I} \varphi_1(f^i(x)) + \sum_{j=1}^{l} \varphi_{m_j - n_j}(f^{n_j}(x)) \tag{3.5}$$

where $I = \bigcup_{j=1}^{l} [m_{j-1}, n_j) \cup [m_l, n)$. Observe that

$$\varphi_1(f^i(x)) = \psi_k(f^i(x)) \quad \text{for any} \quad i \in \bigcup_{j=1}^{l} [m_{j-1}, n_j) \cup [m_l, \min\{n_{l+1}, n\}),$$

since $f^i(x) \in E_k^c$ in all these cases. Moreover, as φ_- is constant on orbits (Exercise 3.2) and $\psi_k \geq \varphi_- + \varepsilon$, the relation (3.4) implies that

$$\varphi_{m_j - n_j}(f^{n_j}(x)) \leq \sum_{i=n_j}^{m_j - 1} (\varphi_-(f^i(x)) + \varepsilon) \leq \sum_{i=n_j}^{m_j - 1} \psi_k(f^i(x))$$

for every $j = 1, \ldots, l$. Thus, using (3.5) we conclude that

$$\varphi_n(x) \leq \sum_{i=0}^{\min\{n_{l+1}, n\} - 1} \psi_k(f^i(x)) + \sum_{i=n_{l+1}}^{n-1} \varphi_1(f^i(x)).$$

Since $n_{l+1} > n - k$, the lemma is proved. □

3.1.3 Estimating φ_-

In order to prove (3.2), in this section we prove the following lemma:

Lemma 3.6 $\int \varphi_- \, d\mu = L.$

Proof Suppose, for the time being, that φ_n/n is uniformly bounded from below; that is, there exists $\kappa > 0$ such that $\varphi_n/n \geq -\kappa$ for every n. In particular, $\varphi_- \geq -\kappa > -\infty$. Applying Fatou's lemma to the sequence of non-negative functions $\varphi_n/n + \kappa$, we obtain that φ_- is integrable and

$$\int \varphi_- \, d\mu \leq \lim \int \frac{\varphi_n}{n} \, d\mu = L.$$

To prove the opposite inequality, observe that Lemma 3.5 implies that

$$\frac{1}{n} \int \varphi_n \, d\mu \leq \frac{n-k}{n} \int \psi_k \, d\mu + \frac{k}{n} \int \max\{\psi_k, \varphi_1\} \, d\mu \tag{3.6}$$

Note that $\max\{\psi_k, \varphi_1\} \leq \max\{\varphi_- + \varepsilon, \varphi_1^+\}$ and this last function is integrable. So, the $\limsup_n$ of the final term in (3.6) is non-positive. Hence, taking $n \to \infty$ we obtain that $L \leq \int \psi_k \, d\mu$ for any k. Then, making $k \to \infty$ we conclude that

$$L \leq \int \varphi_- \, d\mu + \varepsilon$$

Finally, making $\varepsilon \to 0$ we obtain that $L \leq \int \varphi_- \, d\mu$. This proves the lemma when φ_n/n is uniformly bounded from below.

Now, let us remove that hypothesis. Define, for each $\kappa > 0$,

$$\varphi_n^\kappa = \max\{\varphi_n, -\kappa n\} \quad \text{and} \quad \varphi_-^\kappa = \max\{\varphi_-, -\kappa\}.$$

The sequence $(\varphi_n^\kappa)_n$ satisfies the hypotheses of Theorem 3.3: it is subadditive and the positive part of φ_1^κ is integrable. Moreover, $\varphi_-^\kappa = \liminf_n (1/n)\varphi_n^\kappa$. Hence, the argument in the previous paragraph shows that

$$\int \varphi_-^\kappa \, d\mu = \inf_n \frac{1}{n} \int \varphi_n^\kappa \, d\mu. \tag{3.7}$$

By the monotone convergence theorem, we also have that

$$\int \varphi_n \, d\mu = \inf_\kappa \int \varphi_n^\kappa \, d\mu \quad \text{and} \quad \int \varphi_- \, d\mu = \inf_\kappa \int \varphi_-^\kappa \, d\mu. \tag{3.8}$$

Combining (3.7) and (3.8), we find that

$$\int \varphi_- \, d\mu = \inf_\kappa \int \varphi_-^\kappa = \inf_\kappa \inf_n \frac{1}{n} \int \varphi_n^\kappa \, d\mu = \inf_n \frac{1}{n} \int \varphi_n \, d\mu = L.$$

This completes the proof of this lemma. □

3.1.4 Bounding φ_+ from above

We are going to show that $\int \varphi_+ \, d\mu \le L$ if every φ_n is bounded from $-\infty$. That will complete the proof of (3.2). First, we prove a couple of auxiliary results:

Lemma 3.7 *If $\phi : M \to \mathbb{R}$ is integrable with respect to μ then*

$$\lim_n \frac{1}{n} \phi(f^n(x)) = 0 \quad \textit{for } \mu\textit{-almost all } x \in M.$$

Proof Fix any $\varepsilon > 0$. Since μ is invariant under f,

$$\begin{aligned} \mu\big(\{x \in M : |\phi(f^n(x))| \ge n\varepsilon\}\big) &= \mu\big(\{x \in M : |\phi(x)| \ge n\varepsilon\}\big) \\ &= \sum_{k=n}^{\infty} \mu\Big(\Big\{x \in M : k \le \frac{|\phi(x)|}{\varepsilon} < k+1\Big\}\Big). \end{aligned}$$

Adding these inequalities over every $n \in \mathbb{N}$, we obtain

$$\begin{aligned} \sum_{n=1}^{\infty} \mu\big(\{x \in M : |\phi(f^n(x))| \ge n\varepsilon\}\big) &= \sum_{k=1}^{\infty} k\mu\Big(\Big\{x \in M : k \le \frac{|\phi(x)|}{\varepsilon} < k+1\Big\}\Big) \\ &\le \int \frac{|\phi|}{\varepsilon} \, d\mu. \end{aligned}$$

Since ϕ is assumed to be integrable, the right-hand side is finite. Hence, we may use the Borel–Cantelli lemma to conclude that the set $B(\varepsilon)$ of points x such that $|\phi(f^n(x))| \ge n\varepsilon$ for infinitely many values of n has measure zero. By the definition of $B(\varepsilon)$, for every $x \notin B(\varepsilon)$ there exists $p \ge 1$ such that $|\phi(f^n(x))| < n\varepsilon$ for every $n \ge p$. Now, consider $B = \bigcup_{i=1}^{\infty} B(1/i)$. Then B has measure zero and $\lim_n (1/n)\phi(f^n(x)) = 0$ for every $x \notin B$. □

Lemma 3.8 *For any fixed k,*

$$\limsup_n \frac{\varphi_{kn}}{n} = k \limsup_n \frac{\varphi_n}{n}.$$

Proof The inequality $\le$ is clear, since φ_{kn}/kn is a subsequence of φ_n/n. To get the other inequality, write $n = kq_n + r_n$ with $r_n \in \{1, \dots, k\}$. By subadditivity,

$$\varphi_n \le \varphi_{kq_n} + \varphi_{r_n} \circ f^{kq_n} \le \varphi_{kq_n} + \psi \circ f^{kq_n}$$

where $\psi = \max\{\varphi_1^+, \dots, \varphi_k^+\}$. Note that $n/q_n \to k$ as $n \to \infty$. Moreover, since $\psi \in L^1(\mu)$, we may use Lemma 3.7 to check that $\psi \circ f^n/n$ converges to zero at μ-almost every point. Thus, dividing the previous relation by n and taking the lim sup when $n \to \infty$, we obtain that

$$\limsup_n \frac{1}{n} \varphi_n \le \limsup_n \frac{1}{n} \varphi_{kq_n} + \limsup_n \frac{1}{n} \psi \circ f^{kq_n} = \frac{1}{k} \limsup_q \frac{1}{q} \varphi_{kq},$$

as claimed in the lemma. □

Lemma 3.9 *Suppose that* $\inf \varphi_n > -\infty$ *for any* n. *Then* $\int \varphi_+ \, d\mu \leq L$.

Proof For each fixed k and $n \geq 1$, consider $\theta_n = -\sum_{j=0}^{n-1} \varphi_k \circ f^{jk}$. Observe that

$$\int \theta_n \, d\mu = -n \int \varphi_k \, d\mu \quad \text{for every } n, \tag{3.9}$$

since f^k preserves the measure μ. As the sequence $(\varphi_n)_n$ is subadditive, we have that $\theta_n \leq -\varphi_{kn}$ for any n. Then, using Lemma 3.8,

$$\theta_- = \liminf_n \frac{\theta_n}{n} \leq -\limsup_n \frac{\varphi_{kn}}{n} = -k \limsup_n \frac{\varphi_n}{n} = -k\varphi_+$$

and so

$$\int \theta_- \, d\mu \leq -k \int \varphi_+ \, d\mu. \tag{3.10}$$

Observe also that the sequence $(\theta_n)_n$ is additive: $\theta_{m+n} = \theta_m + \theta_n \circ f^{km}$ for any m, $n \geq 1$. As $\theta_1 = -\varphi_k$ is bounded from above by $-\inf \varphi_k$, we also have that the function θ_1^+ is bounded and, consequently, integrable. Thus, we can apply Lemma 3.6, together with (3.9), to conclude that

$$\int \theta_- \, d\mu = \lim_n \int \frac{\theta_n}{n} \, d\mu = -\int \varphi_k \, d\mu. \tag{3.11}$$

Putting the relations (3.10) and (3.11) together, we obtain that

$$\int \varphi_+ \, d\mu \leq \frac{1}{k} \int \varphi_k \, d\mu.$$

Finally, taking the infimum on k yields $\int \varphi_+ \, d\mu \leq L$. □

Lemmas 3.6 and 3.9 prove the relation (3.2) and, thus, Theorem 3.3 when $\inf \varphi_n > -\infty$ for any n. In the general case, define

$$\varphi_n^\kappa = \max\{\varphi_n, -\kappa n\} \quad , \quad \varphi_-^\kappa = \max\{\varphi_-, -\kappa\} \quad \text{and} \quad \varphi_+^\kappa = \max\{\varphi_+, -\kappa\}$$

for any constant $\kappa > 0$. The previous arguments can be applied to the sequence $(\varphi_n^\kappa)_n$ for any fixed $\kappa > 0$. Therefore, $\varphi_+^\kappa = \varphi_-^\kappa$ at μ-almost every point and for any $\kappa > 0$. Since $\varphi_-^\kappa \to \varphi_-$ and $\varphi_+^\kappa \to \varphi_+$ when $\kappa \to \infty$, it follows that $\varphi_- = \varphi_+$ at μ-almost every point. The proof of Theorem 3.3 is complete.

Corollary 3.10 (Birkhoff ergodic theorem) *Let* $\varphi : M \to \mathbb{R}$ *be a* μ*-integrable function. Then*

$$\tilde{\varphi}(x) = \lim_n \frac{1}{n} \sum_{j=0}^{n-1} \varphi(f^j(x))$$

exists at μ*-every point. Moreover, the function* $\tilde{\varphi}$ *is invariant and* μ*-integrable, with* $\int \tilde{\varphi} \, d\mu = \int \varphi \, d\mu$.

Proof According to Example 3.1, this is a particular case of Theorem 3.3. □

Now we can also deduce the following strong version of Lemma 3.7 (the conclusion is the same while the assumption is weaker) that will be useful later.

Corollary 3.11 *Let $\phi : M \to \mathbb{R}$ be a measurable function such that the function $\psi = \phi \circ f - \phi$ is integrable with respect to μ. Then*

$$\lim_n \frac{1}{n}\phi(f^n(x)) = 0 \quad \textit{for } \mu\textit{-almost all } x \in M.$$

In particular, this holds if $\phi \in L^1(\mu)$.

Proof Note that $\phi(f^n(x)) = \phi(x) + \sum_{j=0}^{n-1} \psi(f^j(x))$ for every x and every n. So, by the Birkhoff ergodic theorem applied to the integrable function ψ,

$$\lim_n \frac{1}{n}\phi(f^n(x)) = \lim_n \frac{1}{n}\phi(x) + \lim_n \frac{1}{n}\sum_{j=0}^{n-1} \psi(f^j(x)) \tag{3.12}$$

exists at μ-almost every point. On the other hand, since μ is f-invariant,

$$\mu\big(\{x : |\frac{1}{n}\phi(f^n(x))| \geq c\}\big) = \mu\big(\{y : |\phi(y)| > nc\}\big) \to 0 \quad \text{when } n \to \infty.$$

In other words, the sequence $(1/n)\,\phi \circ f^n$ converges to zero in measure. Thus, the limit in (3.12) must be zero at μ-almost every point. □

3.2 Theorem of Furstenberg and Kesten

Let $F : M \times \mathbb{R}^d \to M \times \mathbb{R}^d$ be given by $F(x,v) = (f(x), A(x)v)$, for some measurable function $A : M \to \mathrm{GL}(d)$. Let $L^1(\mu)$ denote the space of μ-integrable functions on M.

Theorem 3.12 (Furstenberg–Kesten) *If $\log^+ \|A^{\pm 1}\| \in L^1(\mu)$ then*

$$\lambda_+(x) = \lim_n \frac{1}{n} \log \|A^n(x)\| \quad \textit{and} \quad \lambda_-(x) = \lim_n \frac{1}{n} \log \|(A^n(x))^{-1}\|^{-1}$$

exist for μ-almost everywhere $x \in M$. Moreover, the functions $\lambda_\pm$ are invariant and μ-integrable, with

$$\int \lambda_+ d\mu = \lim_n \frac{1}{n} \int \log \|A^n(x)\| d\mu$$

$$\int \lambda_- d\mu = \lim_n \frac{1}{n} \int \log \|(A^n(x))^{-1}\|^{-1} d\mu.$$

Proof This is a direct consequence of Theorem 3.3. Define

$$\varphi_n(x) = \log \|A^n(x)\| \quad \text{and} \quad \psi_n(x) = \log \|(A^n(x))^{-1}\|$$

The hypothesis implies that $\varphi_1^+, \psi_1^+ \in L^1(\mu)$ and so $\varphi_1(x), \psi_1(x) \in [-\infty, +\infty)$ for μ-almost every x. Since the norm of linear operators is sub-multiplicative, the sequences φ_n and ψ_n are subadditive (Example 3.2). Now the conclusion of Theorem 3.12 follows immediately from applying Theorem 3.3 to these sequences. □

3.3 Herman's formula

Michael Herman devised a method for bounding Lyapunov exponents from below, based on the theory of (sub)harmonic functions. Here is one application.

Theorem 3.13 (Herman's formula) *Let $S^1 = \mathbb{R}/\mathbb{Z}$ and $f : S^1 \to S^1$ be an irrational rotation. Let $A : S^1 \to \mathrm{SL}(2)$ be of the form $A(x) = A_0 R_{2\pi x}$ where $A_0 = \begin{pmatrix} \sigma & 0 \\ 0 & \sigma^{-1} \end{pmatrix}$ for some $\sigma > 0$ and $R_\theta = \begin{pmatrix} \cos\theta & -\sin\theta \\ \sin\theta & \cos\theta \end{pmatrix}$. Then*

$$\lambda_+ \geq \log \frac{\sigma + \sigma^{-1}}{2} \tag{3.13}$$

where λ_+ is the largest Lyapunov exponent of the cocycle defined by A, relative to the unique f-invariant probability measure m.

Proof Let $\omega \in 2\pi(\mathbb{R} \setminus \mathbb{Q})$ be the angle of the rotation $f : S^1 \to S^1$. By Theorem 3.12,

$$\begin{aligned} \lambda_+ &= \lim_n \frac{1}{n} \int_{S^1} \log \|A^n(y)\| \, dm(y) \\ &= \lim_n \frac{1}{2\pi n} \int_0^{2\pi} \log \|A_0 R_{x+(n-1)\omega} \cdots A_0 R_{x+\omega} A_0 R_x\| \, dx. \end{aligned} \tag{3.14}$$

This does not depend on the choice of the norm; we take $\|\cdot\|$ to be given by the maximum absolute value of the coefficients. The idea of the proof is to extend the function in the last integral to a subharmonic function on the complex plane and then deduce the claim of the theorem from the average property of subharmonic functions (the value at any point is not bigger than the average of the function over any circle centered at that point). For this, consider the complex matrices

$$\mathscr{R}(z) = \begin{pmatrix} (z^2+1)/2 & (-z^2+1)/(2i) \\ (z^2-1)/(2i) & (z^2+1)/2 \end{pmatrix}$$

defined for $z \in \mathbb{C}$. Observe that $\mathscr{R}(e^{i\theta}) = e^{i\theta} R_\theta$ for every θ. Thus, $z \mapsto \mathscr{R}(z)$ is a kind of holomorphic extension of the family $\theta \mapsto R_\theta$ of rotations. Now write

$$C_n(z) = A_0 \mathscr{R}(e^{(n-1)\omega i} z) \cdots A_0 \mathscr{R}(e^{\omega i} z) A_0 \mathscr{R}(z)$$

Then $C_n(e^{ix}) = e^{i\tau} A^n(x)$ with $\tau = nx + n(n-1)\omega/2$, and so (3.14) becomes

$$\lambda_+ = \lim_n \frac{1}{2\pi n} \int_0^{2\pi} \log \|C_n(e^{ix})\| \, dx.$$

The function $z \mapsto \log \|C_n(z)\|$ is subharmonic because the absolute value of a holomorphic function is subharmonic and the maximum of subharmonic functions is subharmonic (see [45, § 19.4]). It follows that

$$\lambda_+ \geq \lim_n \frac{1}{n} \log \|C_n(0)\|.$$

Moreover,

$$\lim_n \frac{1}{n} \log \|C_n(0)\| = \lim_n \frac{1}{n} \log \|(A_0 \mathscr{R}(0))^n\|$$

equals the logarithm of the spectral radius of $A_0 \mathscr{R}(0)$. An explicit calculation gives that the spectral radius is $(\sigma + \sigma^{-1})/2$. This proves the claim. □

For $\sigma \neq 1$ the theorem gives that λ_+ is positive. Observe that $\deg(f) = 1$ and $\deg(\theta) = 1$, where $\theta(x) = x$ for $x \in S^1$. So, by the criterion in Example 2.9, the cocycle cannot be hyperbolic.

3.4 Theorem of Oseledets in dimension 2

We are going to state and prove a version of the multiplicative ergodic theorem for 2-dimensional cocycles, both invertible and non-invertible. The full statement of the theorem will the topic of Chapter 4. The reason for including this discussion of a special case is that it can handled by much simpler methods and, thus, provides a useful glimpse into the main features of the theorem, in a very transparent situation.

Let $F : M \times \mathbb{R}^2 \to M \times \mathbb{R}^2$ be given by $F(x,v) = (f(x), A(x)v)$, for some measurable function $A : M \to \mathrm{GL}(2)$ satisfying $\log^+ \|A^{\pm 1}\| \in L^1(\mu)$.

3.4.1 One-sided theorem

Theorem 3.14 *For μ-almost every $x \in M$,*

(1) *either* $\lambda_-(x) = \lambda_+(x)$ *and*

$$\lim_n \frac{1}{n} \log \|A^n(x)v\| = \lambda_\pm(x), \textit{ for all } v \in \mathbb{R}^2;$$

(2) *or* $\lambda_+(x) > \lambda_-(x)$ *and there exists a vector line* $E^s_x \subset \mathbb{R}^2$ *such that*

$$\lim_n \frac{1}{n} \log \|A^n(x)v\| = \begin{cases} \lambda_-(x) & \textit{if } v \in E^s_x \setminus \{0\} \\ \lambda_+(x) & \textit{if } v \in \mathbb{R}^d \setminus E^s_x. \end{cases}$$

Moreover $A(x)E^s_x = E^s_{f(x)}$ *for every* x *as in* (2).

Proof We treat the case when A takes values in SL(2) and leave it to the reader (Exercise 3.3) to extend the conclusions to the general GL(2) setting. Consider any x as in the conclusion of Theorem 3.12 and let $\lambda(x) = \lambda_+(x) = -\lambda_-(x)$.

First, let us consider $x \in M$ such that $\lambda(x) = 0$. For any $v \in \mathbb{R}^2$,

$$\|A^n(x)\|^{-1}\|v\| = \|A^n(x)^{-1}\|^{-1}\|v\| \leq \|A^n(x)v\| \leq \|A^n(x)\|\|v\|,$$

and so

$$\frac{1}{n}\log(\|A^n(x)\|^{-1}\|v\|) \leq \frac{1}{n}\log\|A^n(x)v\| \leq \frac{1}{n}\log(\|A^n(x)\|\|v\|).$$

Sending $n \to \infty$ the left-hand side goes to $-\lambda(x) = 0$ and the right-hand side goes to $\lambda(x) = 0$. So, we are done with the case when the exponent vanishes.

Now let us suppose that $\lambda(x) > 0$. Then $\|A^n(x)\| \approx e^{n\lambda(x)}$ is larger than 1 for every large n. So (Exercise 2.3), there exist unit vectors $s_n(x)$ and $u_n(x)$, respectively, most contracted and most expanded under $A^n(x)$:

$$\|A^n(x)s_n(x)\| = \|A^n(x)\|^{-1} \quad \text{and} \quad \|A^n(x)u_n(x)\| = \|A^n(x)\|. \tag{3.15}$$

Lemma 3.15 *The angle* $\measuredangle(s_n(x), s_{n+1}(x))$ *decreases exponentially:*

$$\limsup_n \frac{1}{n} \log|\sin\measuredangle(s_n(x), s_{n+1}(x))| \leq -2\lambda(x).$$

Proof Let us denote $\alpha_n = \measuredangle(s_n(x), s_{n+1}(x))$. Then

$$s_n(x) = \sin\alpha_n u_{n+1}(x) + \cos\alpha_n s_{n+1}(x).$$

It follows that

$$\|A^{n+1}(x)s_n(x)\| \geq \|\sin\alpha_n A^{n+1}(x)u_{n+1}(x)\| = |\sin\alpha_n|\,\|A^{n+1}(x)\|,$$

and

$$\|A^{n+1}(x)s_n(x)\| \leq \|A(f^n(x))\|\|A^n(x)s_n(x)\| = \|A(f^n(x))\|\|A^n(x)\|^{-1}.$$

Hence

$$|\sin \alpha_n| \leq \frac{\|A(f^n(x))\|}{\|A^{n+1}(x)\| \, \|A^n(x)\|}.$$

Corollary 3.11 ensures that

$$\lim_n \frac{1}{n} \log \|A(f^n(x))\| = 0.$$

So, taking limit superior on both sides of the previous expression, we find that

$$\limsup_n \frac{1}{n} \log |\sin \alpha_n| \leq -2\lambda(x).$$

This completes the proof of the lemma. □

Lemma 3.16 *The sequence $(s_n(x))_n$ is Cauchy in projective space.*

Proof Consider any $\varepsilon > 0$ such that $-2\lambda(x) + \varepsilon < 0$. By Lemma 3.15,

$$|\sin \alpha_n| \leq e^{n(-2\lambda(x)+\varepsilon)}$$

for every large n. Then, up to replacing some $s_j(x)$ by $-s_j(x)$,

$$\|s_n(x) - s_{n+1}(x)\| \leq 2e^{n(-2\lambda(x)+\varepsilon)}$$

for every large n. Consequently, there exists $C > 0$ such that

$$\|s_{n+k}(x) - s_n(x)\| \leq Ce^{n(-2\lambda(x)+\varepsilon)}$$

for every $k \geq 1$ and n large enough. In particular, the sequence is Cauchy. □

Define $s(x) = \lim s_n(x)$ whenever the limit exists.

Lemma 3.17 *The vector $s(x)$ is contracted at the rate $-\lambda(x)$:*

$$\lim_n \frac{1}{n} \log \|A^n(x)s(x)\| = -\lambda(x).$$

Proof Let $\beta_n = \measuredangle(s(x), s_n(x))$. Then $s(x) = \cos \beta_n s_n(x) + \sin \beta_n u_n(x)$ and so

$$\|A^n(x)s(x)\| \leq |\cos \beta_n| \|A^n(x)s_n(x)\| + |\sin \beta_n| \|A^n(x)u_n(x)\|.$$

Then (Exercise 3.6),

$$\begin{aligned}
&\limsup_n \frac{1}{n} \log \|A^n(x)s(x)\| \\
&\quad \le \max\left\{\limsup_n \frac{1}{n} |\cos\beta_n|\, \|A^n(x)s_n(x)\|, \limsup_n \frac{1}{n} |\sin\beta_n|\, \|A^n(x)u_n(x)\|\right\} \\
&\quad \le \max\left\{\limsup_n \frac{1}{n} \log \|A^n(x)\|^{-1},\right. \\
&\qquad\qquad \left.\limsup_n \frac{1}{n} \log|\sin\beta_n| + \limsup_{n\to\infty} \frac{1}{n} \|A^n(x)u_n(x)\|\right\} \\
&\quad \le \max\{-\lambda(x), -2\lambda(x)+\lambda(x)\} = -\lambda(x).
\end{aligned}$$

This completes the proof. □

Lemma 3.18 *If $v \in \mathbb{R}^2$ is not collinear with $s(x)$ then*

$$\lim_n \frac{1}{n} \log \|A^n(x)v\| = \lambda(x).$$

Proof Denote $\gamma_n = \measuredangle(v, s_n(x))$. Then $v = \cos\gamma_n s_n(x) + \sin\gamma_n u_n(x)$ and so

$$\|A^n(x)v\| \ge |\sin\gamma_n| \|A^n(x)u_n(x)\| - |\cos\gamma_n| \|A^n(x)s_n(x)\|.$$

Note that $|\sin\gamma_n|$ is bounded from zero for all large n, because $s_n(x) \to s(x)$ and v is not collinear to $s(x)$. Using also (3.15),

$$\|A^n(x)u_n(x)\| \approx e^{n\lambda(x)} \quad \text{and} \quad \|A^n(x)s_n(x)\| \approx e^{-n\lambda(x)}.$$

Substituting this in the previous inequality, and taking the limit as $n \to \infty$, we get

$$\liminf_n \frac{1}{n} \log \|A^n(x)v\| \ge \lambda(x).$$

From $\|A^n(x)v\| \le \|A^n(x)\| \|v\|$ we immediately get the opposite inequality:

$$\limsup_n \frac{1}{n} \log \|A^n(x)v\| \le \lim_n \frac{1}{n} \log \|A^n(x)\| = \lambda(x).$$

The proof of the lemma is complete. □

Lemma 3.19 *$A(x)s(x)$ is collinear to $s(f(x))$.*

Proof By Lemma 3.17,

$$\lim_n \frac{1}{n} \log \|A^n(f(x))A(x)s(x)\| = \lim_n \frac{1}{n+1} \log \|A^{n+1}(x)s(x)\| = -\lambda(x).$$

By Lemma 3.18,

$$\lim_n \frac{1}{n} \log \|A^n(f(x))v\| = \lambda(f(x)) = \lambda(x)$$

for every v not collinear to $s(f(x))$. This implies the claim of the lemma. □

Take E^s_x to be the line $\mathbb{R}s(x)$ generated by $s(x)$. Lemmas 3.15 through 3.19 contain all the claims in Theorem 3.14. □

3.4.2 Two-sided theorem

Theorem 3.20 *If $f : M \to M$ is invertible then for μ-almost every $x \in M$:*

(1) *either $\lambda_-(x) = \lambda_+(x)$ and*

$$\lim_{n\to\pm\infty} \frac{1}{n} \log \|A^n(x)v\| = \lambda_\pm(x) \text{ for all } v \in \mathbb{R}^2;$$

(2) *or $\lambda_-(x) < \lambda_+(x)$ and there is a direct sum decomposition $\mathbb{R}^2 = E^u_x \oplus E^s_x$ such that*

$$\lim_{n\to+\infty} \frac{1}{n} \log \|A^n(x)v\| = \begin{cases} \lambda_-(x) & \text{if } v \in E^s_x \setminus \{0\} \\ \lambda_+(x) & \text{if } v \in \mathbb{R}^d \setminus E^s_x \end{cases}$$

$$\lim_{n\to-\infty} \frac{1}{n} \log \|A^n(x)v\| = \begin{cases} \lambda_+(x) & \text{if } v \in E^u_x \setminus \{0\} \\ \lambda_-(x) & \text{if } v \in \mathbb{R}^d \setminus E^u_x. \end{cases}$$

Moreover, in the latter case, $A(x)E^u_x = E^u_{f(x)}$ and $A(x)E^s_x = E^s_{f(x)}$ and the angle between the two lines decreases subexponentially along orbits:

$$\lim_{n\to\pm\infty} \frac{1}{n} \log |\sin \measuredangle(E^u_{f^n(x)}, E^s_{f^n(x)})| = 0.$$

Proof We deal with the case when $A(x) \in \mathrm{SL}(2)$ for all x and let the reader extend the conclusions to the general setting. For x as in the conclusion of Theorem 3.12, write $\lambda(x) = \lambda_+(x) = -\lambda_-(x)$.

The case $\lambda(x) = 0$ follows directly from Theorem 3.14 applied to F and to its inverse F^{-1}. From now on, assume that $\lambda(x) > 0$. Let $E^s_x = \mathbb{R}s(x)$ and $E^u_x = \mathbb{R}u(x)$ be the subspaces given by Theorem 3.14 for F and F^{-1}, respectively. We need to check that these two lines are transverse:

Lemma 3.21 *The vectors $s(x)$ and $u(x)$ are non-collinear, for μ-almost every point in $\{x : \lambda(x) > 0\}$.*

Proof We have $\lim_{n\to-\infty} n^{-1} \log \|A^n(x) \mid E^u_x\| = \lambda(x)$, by Theorem 3.14 applied to F^{-1}. Thus, the lemma will follow if we prove that

$$\lim_{n\to-\infty} \frac{1}{n} \log \|A^n(x) \mid E^s_x\| = -\lambda(x).$$

From Theorem 3.14 applied to F^{-1}, we know that the limit on the left-hand side exists. Let us denote it by $\psi(x)$. Consider the sequence of functions

$$\psi_n(x) = \frac{1}{-n}\log\|A^{-n}(x) \mid E_x^s\| \quad \text{and} \quad \phi_n(y) = \frac{1}{-n}\log\|(A^n(y) \mid E_y^s)^{-1}\|.$$

From the definition of A^{-n} we get that $\psi_n(x) = \phi_n(f^{-n}(x))$ for every $n \geq 1$. Since E^s is one-dimensional, we may write

$$\phi_n(x) = \frac{1}{n}\log\|A^n(y) \mid E_y^s\|.$$

Then $\lim_{n\to\infty}\phi_n(y) = -\lambda(y)$, by Lemma 3.17. In particular, the sequence ϕ_n converges to $-\lambda$ in measure; that is,

$$\lim_n \mu(\{y : |\phi_n(y) + \lambda_-(y)| > \delta\}) = 0 \quad \text{for any } \delta > 0.$$

Then, since μ is f-invariant,

$$\lim_n \mu(\{y : |\phi_n(f^{-n}(x)) + \lambda_-(f^{-n}(x))| > \delta\}) = 0 \quad \text{for any } \delta > 0.$$

In view of the previous observations, and the fact that the function λ_- is invariant, this implies that

$$\lim_n \mu(\{y : |\psi_n(x) + \lambda(x)| > \delta\}) = 0 \quad \text{for any } \delta > 0.$$

This means that ψ_n converges to $-\lambda$ in measure. On the other hand, ψ_n converges to ψ almost everywhere and, consequently, in measure. By uniqueness of the limit, it follows that $\psi = \lambda_-$, as claimed. □

Lemma 3.22 *Let $\theta(y) = \measuredangle(E_y^s, E_y^u)$. For μ-almost every x with $\lambda(x) > 0$,*

$$\lim_{n\to\pm\infty}\frac{1}{n}\log|\sin\theta(f^n(x))| = 0.$$

Proof From the elementary relation (Exercise 3.4)

$$\|A(x)\|^{-2} \leq \frac{|\sin\theta(f(x))|}{|\sin\theta(x)|} \leq \|A(x)\|^2$$

we find that $\big|\log|\sin\theta(f(x))| - \log|\sin\theta(x)|\big| \leq 2\log\|A(x)\|$ and, in particular, $\log|\sin\theta|\circ f - \log|\sin\theta| \in L^1(\mu)$. So we may conclude from Corollary 3.11 that

$$\lim_{n\to\pm\infty}\frac{1}{n}\log|\sin\theta(f^n(x))| = 0$$

as stated. □

This finishes the proof of Theorem 3.20. □

3.5 Notes

Theorem 3.3 is an application of the main result of Kingman [74]. The proof we presented here is due to Avila and Bochi [10], inspired by a proof of the Birkhoff ergodic theorem by Katznelson and Weiss [69]. See Ledrappier [80, § I.2] for another proof inspired by [69].

Theorem 3.12, which we obtained as a consequence of the subadditive ergodic theorem, was actually proven before, by Furstenberg and Kesten [56]. The subadditive ergodic theorem will reappear in Chapter 4, as part of the proof of the Oseledets theorem.

Theorem 3.13 is perhaps the simplest application of the method devised by Herman [62] for estimating Lyapunov exponents from below. Avila and Bochi [11] proved that (3.13) is, actually, an equality.

The arguments in Section 3.4 are inspired by the proof of Proposition 2.1. Other proofs of the Oseledets theorem in two dimensions have been given by Young [119] and Bochi [29].

3.6 Exercises

Exercise 3.1 Check that, given any measurable function $A : M \to \mathrm{GL}(d)$,

$$\log^+ \|A^{\pm 1}\| \in L^1(\mu) \Leftrightarrow |\log \|A^{\pm 1}\|| \in L^1(\mu) \Leftrightarrow \log^- \|A^{\pm 1}\| \in L^1(\mu).$$

Moreover, if A takes values in $\mathrm{SL}(d)$, all these conditions are equivalent to $\log \|A\| \in L^1(\mu)$.

Exercise 3.2 Check that the functions φ_- and φ_+ are invariant.

Exercise 3.3 Given $A : M \to \mathrm{GL}(d)$, define $c(x) = |\det A(x)|^{1/d}$ and then let $B : M \to \mathrm{SL}(d)$ be given by $A(x) = c(x)B(x)$. Show that $\log c$ and $\log^+ \|B^{\pm 1}\|$ are in $L^1(\mu)$ if $\log^+ \|A^{\pm 1}\|$ is in $L^1(\mu)$. Check that, for μ-almost every $x \in M$,

$$\lim_n \frac{1}{n} \log \|A^n(x)v\| = t(x) + \lim_n \frac{1}{n} \log \|B^n(x)v\| \quad \text{for every } v \neq 0,$$

where $t(x) = \lim_n n^{-1} \sum_{j=0}^{n-1} \log c(f^j(x))$ is the Birkhoff time average of the function $\log c$. Deduce that the associated cocycles $F(x,v) = (f(x), A(x)v)$ and $G(x,v) = (f(x), B(x)v)$ have the same Oseledets flag/decomposition at almost every point. Moreover, the Lyapunov spectrum of the former cocycle is the $t(x)$-translate of the Lyapunov spectrum of the latter.

Exercise 3.4 Show that, for any $B \in \mathrm{SL}(2)$ and non-zero vectors u, $v \in \mathbb{R}^2$,

$$\|B\|^{-2} \le \frac{|\sin\measuredangle(B(u),B(v))|}{|\sin\measuredangle(u,v)|} \le \|B\|^2.$$

Exercise 3.5 Prove that $\lim_{n\to\infty} n^{-1}\log|\sin\measuredangle(A^n(x)v, E^u_{f^n(x)})| = -2\lambda(x)$ for any $v \notin E^s_x$. If the cocycle F is invertible then the same holds as $n \to -\infty$, with E^u_x in the place of E^s_x.

Exercise 3.6 Show that if $a_n, b_n > 0$ for every n then

(1) $\displaystyle\limsup_n \frac{1}{n}\log(a_n+b_n) = \max\left\{\limsup_n \frac{1}{n}\log a_n, \limsup_n \frac{1}{n}\log b_n\right\}.$

(2) $\displaystyle\limsup_n \frac{1}{n}\log\sqrt{a_n^2+b_n^2} = \max\left\{\limsup_n \frac{1}{n}\log a_n, \limsup_n \frac{1}{n}\log b_n\right\}.$

(3) $\displaystyle\liminf_n \frac{1}{n}\log(a_n+b_n) \ge \max\left\{\liminf_n \frac{1}{n}\log a_n, \liminf_n \frac{1}{n}\log b_n\right\}.$

(4) $\displaystyle\liminf_n \frac{1}{n}\log\sqrt{a_n^2+b_n^2} \ge \max\left\{\liminf_n \frac{1}{n}\log a_n, \liminf_n \frac{1}{n}\log b_n\right\}.$

4
Multiplicative ergodic theorem

This chapter is devoted to the fundamental result in the theory of Lyapunov exponents: the multiplicative ergodic theorem of Oseledets [92]. The statements are given in Section 4.1 and proofs appear in Section 4.2 (one-sided version, for general cocycles) and Section 4.3 (two-sided version, for invertible cocycles). Some related issues are discussed in Section 4.4.

Throughout, we take $(M, \mathscr{B}, \mu)$ to be a complete separable probability space. Recall that *complete* means that any subset of a measurable set with zero measure is measurable (and has zero measure) and *separable* means that there exists a countable family $\mathscr{E} \subset \mathscr{B}$ such that for any $\varepsilon > 0$ and any $B \in \mathscr{B}$ there exists $E \in \mathscr{E}$ such that $\mu(B \Delta E) < \varepsilon$.

Let $F : M \times \mathbb{R}^d \to M \times \mathbb{R}^d$ be the linear cocycle defined by a measurable function $A : M \to \mathrm{GL}(d)$ over a measurable transformation $f : M \to M$ that preserves the probability measure μ. It is assumed that the functions $\log^+ \|A^{\pm 1}\|$ are integrable with respect to μ.

4.1 Statements

Recall that the Grassmannian of $\mathbb{R}^d$ is the disjoint union $\mathrm{Gr}(d)$ of the Grassmannian manifolds $\mathrm{Gr}(l,d)$, $0 \le l \le d$. A map $x \mapsto V_x$ with values in $\mathrm{Gr}(d)$ is measurable if and only if there exist measurable, linearly independent vector fields that span V_x at each point (Exercise 4.1).

A *flag* in $\mathbb{R}^d$ is a decreasing family $W^1 \supsetneq \cdots \supsetneq W^k \supsetneq \{0\}$ of vector subspaces of the d-dimensional Euclidean space. The flag is called *complete* if $k = d$ and $\dim W^j = d + 1 - j$ for all $j = 1, \ldots, d$.

Theorem 4.1 (Oseledets) *For μ-almost every $x \in M$ there is $k = k(x)$, num-*

bers $\lambda_1(x) > \cdots > \lambda_k(x)$ *and a flag* $\mathbb{R}^d = V_x^1 \supsetneq \cdots \supsetneq V_x^k \supsetneq \{0\}$, *such that, for all* $i = 1, \ldots, k$*:*

(a) $k(f(x)) = k(x)$ *and* $\lambda_i(f(x)) = \lambda_i(x)$ *and* $A(x) \cdot V_x^i = V_{f(x)}^i$*;*

(b) *the maps* $x \mapsto k(x)$ *and* $x \mapsto \lambda_i(x)$ *and* $x \mapsto V_x^i$ *(with values in* $\mathbb{N}$ *and* $\mathbb{R}$ *and* $\mathrm{Gr}(d)$*, respectively) are measurable;*

(c) $\lim_n \frac{1}{n} \log \|A^n(x)v\| = \lambda_i(x)$ *for all* $v \in V_x^i \setminus V_x^{i+1}$ *(with* $V_x^{k+1} = \{0\}$*).*

When μ is ergodic, it follows that the values of $k(x)$ and of each of the *Lyapunov exponents* $\lambda_i(x)$ are constant on a full measure subset, and so are the dimensions of the *Oseledets subspaces* V_x^i. We call $\dim V_x^i - \dim V_x^{i+1}$ the *multiplicity* of the corresponding *Lyapunov exponent* $\lambda_i(x)$. The *Lyapunov spectrum* of F is the set of all Lyapunov exponents, each counted with multiplicity. The Lyapunov spectrum is *simple* if all Lyapunov exponents have multiplicity 1 or, equivalently, if the Oseledets flag is complete.

When the transformation $f : M \to M$ is invertible, we have a stronger conclusion:

Theorem 4.2 (Oseledets) *Suppose that* $f : M \to M$ *is invertible. Then, for* μ*-almost every* $x \in M$*, there exists a direct sum decomposition* $\mathbb{R}^d = E_x^1 \oplus \cdots \oplus E_x^k$ *such that, for every* $i = 1, \ldots, k$*:*

(a) $A(x) \cdot E_x^i = E_{f(x)}^i$ *and* $V_x^i = \bigoplus_{j=i}^k E_x^j$ *and*

(b) $\lim_{n \to \pm\infty} \frac{1}{n} \log \|A^n(x)v\| = \lambda_i(x)$ *for all* $v \in E_x^i \setminus \{0\}$ *and*

(c) $\lim_{n \to \pm\infty} \frac{1}{n} \log \left| \sin \measuredangle\left(\bigoplus_{i \in I} E_{f^n(x)}^i, \bigoplus_{j \in J} E_{f^n(x)}^j \right) \right| = 0$ *whenever* $I \cap J = \emptyset$*.*

By definition, the *angle* $\measuredangle(V, W)$ between two subspaces V and W of $\mathbb{R}^d$ is the smallest angle between non-zero vectors $v \in V$ and $w \in W$.

Clearly, the multiplicity of each Lyapunov exponent λ_i coincides with the dimension $\dim E_x^i = \dim V_x^i - \dim V_x^{i+1}$ of the associated *Oseledets subspace* E_x^i. Thus, the Lyapunov spectrum is simple if and only if $\dim E_x^i = 1$ for every i.

Remark 4.3 The sums $\bigoplus_{j=i}^k E_x^j$ of Oseledets subspaces corresponding to the smallest Lyapunov exponents depend only on the forward iterates of the linear cocycle, since they coincide with the subspaces V_x^i. Analogously, any sum of Oseledets subspaces corresponding to the largest Lyapunov exponents depends only on the backward iterates.

4.2 Proof of the one-sided theorem

As we are going to see, it is rather easy to get a weaker form of Theorem 4.1, where limit is replaced by limit superior in part (c). In Section 4.2.1 we find the Lyapunov exponents $\lambda_i(x)$ and the Oseledets subspaces V_x^i and we check that they are invariant. Then, in Section 4.2.2, we show that these objects are measurable.

The main difficulty is to prove that the limit in part (c) actually exists. Roughly speaking, the proof is by induction on the number k of Oseledets subspaces. Sections 4.2.3 and 4.2.4 prepare the proof of the case $k = 1$, whereas the inductive step is prepared in Section 4.2.5. In Section 4.2.6 we wrap up the proof.

4.2.1 Constructing the Oseledets flag

For each $v \in \mathbb{R}^d \setminus \{0\}$ and $x \in M$, define

$$\lambda(x,v) = \limsup_n \frac{1}{n} \log \|A^n(x)v\|. \tag{4.1}$$

In the next lemma we collect a few basic properties of this function. Recall that, by Theorem 3.12, the extremal Lyapunov exponents $\lambda_\pm(x)$ are well defined and finite for μ-almost every x.

Lemma 4.4 *For μ-almost every $x \in M$ and any $v, v' \in \mathbb{R}^d \setminus \{0\}$,*

(i) $\lambda_-(x) \le \lambda(x,v) \le \lambda_+(x)$;
(ii) $\lambda(x,cv) = \lambda(x,v)$ *for* $v \in \mathbb{R}^d \setminus \{0\}$ *and* $c \neq 0$;
(iii) $\lambda(x,v+v') = \max\{\lambda(x,v), \lambda(x,v')\}$ *if* $v+v' \neq 0$;
(iv) $\lambda(x,v) = \lambda(f(x), A(x)v)$.

Proof For claim (i), just observe that

$$\|A^n(x)^{-1}\|^{-1}\|v\| \le \|A^n(x)v\| \le \|A^n(x)\| \, \|v\|$$

and take the limit/limit superior as $n \to \infty$. Part (ii) follows directly from the definition (4.1). Claim (iii) is a direct consequence of Exercise 3.6. Part (iv) also follows directly from the definition. □

For any $x \in M$ as in Lemma 4.4, take $k(x) \ge 1$ to be the number of elements of $\{\lambda(x,v) : v \in \mathbb{R}^d \setminus \{0\}\}$, let $\lambda_1(x) > \cdots > \lambda_{k(x)}(x)$ be those elements, in decreasing order, and let

$$V_x^i = \{v \in \mathbb{R}^d \setminus \{0\} : \lambda(x,v) \le \lambda_i(x)\} \cup \{0\} \quad \text{for } i = 1, \ldots, k(x).$$

By part (i) of Lemma 4.4, every $\lambda_i(x)$ is a real number. Moreover, parts (ii) and (iii) ensure that V_x^i is a vector subspace, for every i.

It follows directly from these definitions that the V_x^i constitute a flag:

$$\mathbb{R}^d = V_x^1 \supsetneq \cdots \supsetneq V_x^{k(x)} \supsetneq \{0\}.$$

In particular, $k(x) \leq d$. Another direct consequence of the definitions is:

$$\lambda(x,v) = \lambda_i(x) \quad \text{for every } v \in V_x^i \setminus V_x^{i+1}. \tag{4.2}$$

In particular, by part (i) of Lemma 4.4,

$$\lambda_-(x) \leq \lambda_i(x) \leq \lambda_+(x) \quad \text{for every } i = 1, \ldots, k(x). \tag{4.3}$$

Finally, by Lemma 4.4(iv), the functions $x \mapsto k(x)$ and $x \mapsto \lambda_i(x)$ and $x \mapsto V_x^i$ are all invariant: for every $i = 1, \ldots, k(x)$,

$$k(f(x)) = k(x) \quad \text{and} \quad \lambda_i(f(x)) = \lambda_i(x) \quad \text{and} \quad V_{f(x)}^i = A(x)V_x^i.$$

This proves part (a) of the theorem.

4.2.2 Measurability

Next, we are going to show that the functions k, λ_i and V^i are measurable, as claimed in part (b) of the theorem. We will use the following criteria for measurability, whose proof can be found in Castaing and Valadier [43].

Let $(X, \mathscr{B}, \mu)$ be a complete probability space and Y be a separable complete metric space. Denote by $\mathscr{B}(Y)$ the Borel σ-algebra of Y.

Proposition 4.5 (Theorem III.23 in [43]) *Let $\mathscr{B} \otimes \mathscr{B}(Y)$ be the product σ-algebra in $X \times Y$ and $\pi : X \times Y \to X$ be the canonical projection. Then $\pi(E) \in \mathscr{B}$ for every $E \in \mathscr{B} \otimes \mathscr{B}(Y)$.*

Proposition 4.6 (Theorem III.30 in [43]) *Let $\mathscr{K}(Y)$ be the space of compact subsets of Y, with the Hausdorff topology. The following are equivalent:*

(i) *a map $x \mapsto K_x$ from X to $\mathscr{K}(Y)$ is measurable;*
(ii) *its graph $\{(x,y) : y \in K_x\}$ is in $\mathscr{B} \otimes \mathscr{B}(Y)$;*
(iii) *$\{x \in X : K_x \cap U \neq \emptyset\} \in \mathscr{B}$ for any open set $U \subset Y$.*

Moreover, any of these conditions implies that there exists a measurable map $\sigma : X \to Y$ such that $\sigma(x) \in K_x$ for every $x \in X$.

Here is a useful application to maps with values in the Grassmannian:

Corollary 4.7 *Let $(X, \mathscr{B}, \mu)$ be a complete probability space and $x \mapsto V_x$ be a map from X to the Grassmannian* $\operatorname{Gr}(d)$. *The following are equivalent:*

(a) *the map $x \mapsto V_x$ is measurable;*
(b) *its graph $\{(x,v) \in X \times \mathbb{R}^d : v \in V_x\}$ is in $\mathscr{B} \otimes \mathscr{B}(\mathbb{R}^d)$.*

Proof For each subspace V of $\mathbb{R}^d$, define $\mathbb{P}V = \{\xi \in \mathbb{P}\mathbb{R}^d : \xi \subset V\}$. Consider the following projectivized versions of conditions (a) and (b):

(a′) the map $x \mapsto \mathbb{P}V_x$, from M to $\mathscr{K}(\mathbb{P}\mathbb{R}^d)$ is measurable;
(b′) its graph $\{(x,\xi) \in M \times \mathbb{P}\mathbb{R}^d : \xi \in \mathbb{P}V_x\}$ is in $\mathscr{B} \otimes \mathscr{B}(\mathbb{P}\mathbb{R}^d)$.

Note that (a′) $\Leftrightarrow$ (b′), by Proposition 4.6. Next, consider the map

$$\mathbb{P} : \mathrm{Gr}(\mathbb{R}^d) \to \mathscr{K}(\mathbb{P}\mathbb{R}^d), \quad V \mapsto \mathbb{P}V.$$

This map is continuous and injective. Since $\mathrm{Gr}(d)$ is compact, it follows that $\mathbb{P}$ is a homeomorphism onto its image. The composition of $\mathbb{P}$ with the map in (a) is, precisely, the map in (a′). Thus, (a) $\Leftrightarrow$ (a′). Similarly,

$$\begin{aligned}\{(x,v) \in M \times \mathbb{R}^d &: v \in V_x\} \\ &= p^{-1}\big(\{(x,\xi) \in M \times \mathbb{P}\mathbb{R}^d : \xi \in \mathbb{P}V_x\}\big) \cup \big(M \times \{0\}\big).\end{aligned}$$

where

$$p : M \times \big(\mathbb{R}^d \setminus \{0\}\big) \to M \times \mathbb{P}\mathbb{R}^d, \quad p(x,v) = (x,[v]).$$

So, since the map p is measurable, (b′) $\Rightarrow$ (b). Next, we prove that (b) $\Rightarrow$ (b′). Let $\pi : M \times \mathbb{R}^d \to M$ be the canonical projection to the first coordinate and let U be any open subset of $\mathbb{P}\mathbb{R}^d$. Observe that

$$\begin{aligned}\{x \in M &: \mathbb{P}V_x \cap U \neq \emptyset\} \\ &= \pi\big(\{(x,v) \in M \times \mathbb{R}^d : v \in V_x\} \cap p^{-1}(M \times U)\big).\end{aligned} \tag{4.4}$$

Assume (b); that is, take $\{(x,v) \in M \times \mathbb{R}^d : v \in V_x\}$ to be in $\mathscr{B} \otimes \mathscr{B}(\mathbb{R}^d)$. Then, using Proposition 4.5, it follows from (4.4) that $\{x \in M : \mathbb{P}V_x \cap U \neq \emptyset\}$ is in $\mathscr{B}$. By Proposition 4.6(iii), this implies the statement in (b′):

$$\{(x,\xi) \in M \times \mathbb{P}\mathbb{R}^d : \xi \in \mathbb{P}V_x\} \text{ is in } \mathscr{B} \otimes \mathscr{B}(\mathbb{P}\mathbb{R}^d),$$

Thus (b) $\Rightarrow$ (b′). This completes the proof that (a) $\Leftrightarrow$ (b). □

Let $e_1, \ldots, e_d$ be an arbitrary basis of $\mathbb{R}^d$. By Exercise 3.6,

$$\lambda_1(x) = \max\{\lambda(x,e_i) : 1 \leq i \leq d\}.$$

Since $(x,v) \mapsto \lambda(x,v)$ is measurable, it follows that $x \mapsto \lambda_1(x)$ is a measurable function and

$$V_*^2 = \{(x,v) \in M \times \mathbb{R}^d \setminus \{0\} : \lambda(x,v) < \lambda_1(x)\}$$

is a measurable subset of $M \times \mathbb{R}^d$. Observe that

$$\pi(V_*^2) = \{x \in M : \lambda(x,v) < \lambda_1(x) \text{ for some } v \in \mathbb{R}^d \setminus \{0\}\}$$
$$= \{x \in M : k(x) \geq 2\}.$$

By Proposition 4.5, this is a measurable subset of M. For $x \in \pi(V_*^2)$, define

$$V_x^2 = \{v \in \mathbb{R}^d : (x,v) \in V^2\} \cup \{0\}.$$

Since $V_*^2 \cup (M \times \{0\})$ is a measurable subset of $M \times \mathbb{R}^d$, Corollary 4.7 gives that $x \mapsto V_x^2$ is a measurable map on $\pi(V_*^2)$. Then, by Exercise 4.1, each

$$M_l^2 = \{x \in \pi(V_*^2) : \dim V_x^2 = l\}, \quad 1 \leq l \leq d$$

is a measurable subset and for each l there exist measurable functions

$$v_1, \ldots, v_l : M_l^2 \to \mathbb{R}^d$$

such that $\{v_1(x), \ldots, v_l(x)\}$ is a basis of V_x^2 for every x. Then

$$\lambda_2(x) = \max\{\lambda(x, u_i(x)) : 1 \leq i \leq l\}$$

is a measurable function on M_l^2, for every $1 \leq l \leq d$. Next, let

$$V_*^3 = \{(x,v) \in M \times \mathbb{R}^d \setminus \{0\} : \lambda(x,v) < \lambda_2(x)\}$$

and $V_x^3 = \{v \in \mathbb{R}^d : (x,v) \in V_*^3\} \cup \{0\}$ for each $x \in \pi(V_*^3)$. Just as before, $\pi(V_*^3) = \{x \in M : k(x) \geq 3\}$ is a measurable subset of M, the map $x \mapsto V_x^3$ is measurable on $\pi(V_*^3)$, each

$$M_l^3 = \{x \in \pi(V_*^3) : \dim V_x^3 = l\}, \quad 1 \leq l \leq d$$

is a measurable subset and, for each l, there exist measurable functions

$$v_1, \ldots, v_l : M_l^3 \to \mathbb{R}^d$$

such that $\{v_1(x), \ldots, v_l(x)\}$ is a basis of V_x^3 for every x. Repeating this argument successively, we find that

(i) $\{x \in M : k(x) \geq s\}$ is measurable for every $s \geq 1$; thus, the function $x \mapsto k(x)$ is measurable;
(ii) each Lyapunov exponent $\lambda_i(x)$ is a measurable function of x on the set $\pi(V_*^i) = \{x \in M : k(x) \geq i\}$;
(iii) each Oseledets subspace V_x^i is a measurable function of x on $\pi(V_*^i)$.

This proves part (b) of the theorem.

4.2.3 Time averages of skew products

The relation (4.2) gives a weaker version of part (c), with limit superior instead of limit. We are left to prove that the limit does exist. The heart of the argument is Proposition 4.8 below.

Let P be a compact metric space. Consider the space $C^0(P)$ of continuous real functions on P, endowed with the norm

$$\|\psi\|_0 = \sup\{|\psi(\xi)| : \xi \in P\}.$$

It is well known that $C^0(P)$ is a separable space (see [114, Theorem A.3.13]).

Denote by $\mathscr{F}$ the space of measurable functions $\Psi : M \times P \to \mathbb{R}$ such that $\Psi(x,\cdot) \in C^0(P)$ for μ-almost every x and $x \mapsto \|\Psi(x,\cdot)\|_0$ is integrable with respect to μ, modulo identifying any two functions that coincide on some $N \times P$ with $\mu(N) = 1$. Then,

$$\|\Psi\|_1 = \int \|\Psi(x,\cdot)\|_0 \, d\mu(x) \tag{4.5}$$

defines a complete norm on $\mathscr{F}$. Using the assumption that $(M, \mathscr{B}, \mu)$ is a separable probability space, one also gets that $(\mathscr{F}, \|\cdot\|)$ is separable. See Exercise 4.2.

Let $\mathscr{M}(\mu)$ be the space of probability measures on $M \times P$ such that $\pi_* \eta = \mu$, where $\pi : M \times P \to M$ is the canonical projection. The *weak* topology* is the smallest topology on $\mathscr{M}(\mu)$ such that the operator

$$\mathscr{M}(\mu) \to \mathbb{R}, \quad \eta \mapsto \int \Psi \, d\eta \quad \text{is continuous, for every } \Psi \in \mathscr{F}.$$

A variation of the classical Banach–Alaoglu argument (see Theorem 2.1 in [114]) shows that this topology is metrizable and compact (Exercise 4.4).

Proposition 4.8 *Let $\mathscr{G} : M \times P \to M \times P$ be a measurable map of the form $\mathscr{G}(x,v) = (f(x), \mathscr{G}_x(v))$, where $\mathscr{G}_x : P \to P$ is a continuous map for μ-almost every x. Given any $\Phi \in \mathscr{F}$, define*

$$I(x) = \lim_n \frac{1}{n} \inf_{v \in P} \sum_{j=0}^{n-1} \Phi(\mathscr{G}^j(x,v)) \quad \text{and} \quad S(x) = \lim_n \frac{1}{n} \sup_{v \in P} \sum_{j=0}^{n-1} \Phi(\mathscr{G}^j(x,v)).$$

The limits exist at μ-almost every point and there exist $\mathscr{G}$-invariant probability measures $\eta_I \in \mathscr{M}(\mu)$ and $\eta_S \in \mathscr{M}(\mu)$ such that

$$\int \Phi \, d\eta_I = \int I \, d\mu \quad \text{and} \quad \int \Phi \, d\eta_S = \int S \, d\mu. \tag{4.6}$$

Proof The roles of the infimum and the supremum are exchanged if one replaces Φ by $-\Phi$. Thus, it suffices to prove the claims pertaining to either one of them. For each $n \geq 1$, define

$$I_n(x) = \inf_{v \in P} \sum_{j=0}^{n-1} \Phi(\mathscr{G}^j(x,v)) \tag{4.7}$$

Every I_n is a measurable function, since it coincides with the infimum taken over any countable dense subset of P. Note that I_1 is integrable, because $\Phi \in \mathscr{F}$. Moreover, the sequence $(I_n)_n$ is *super-additive*:

$$I_{m+n}(x) \geq I_m(x) + I_n(f^m(x)) \quad \text{for every } m, n \text{ and } x.$$

Then, by the subadditive ergodic theorem applied to $(-I_n)_n$,

$$I(x) = \lim_n \frac{1}{n} I_n(x) \quad \text{exists for } \mu\text{-almost every } x.$$

Consider the subsets Γ_n of $M \times P$ defined by

$$\Gamma_n = \{(x,v) \in M \times P : \sum_{j=0}^{n-1} \Phi(\mathscr{G}^j(x,v)) = I_n(x)\}$$

and let $\Gamma_n(x) = \{v \in P : (x,v) \in \Gamma_n\}$ for each $x \in M$. This is a non-empty compact subset of P, for μ-almost every x, since P is compact and $v \mapsto \Phi(\mathscr{G}^j(x,v))$ is continuous for every j. Then, since Γ_n is measurable, Proposition 4.6 gives that there exists a measurable map $v_n : M \to P$ such that $v_n(x) \in \Gamma_n(x)$ for μ-almost every x. Now let

$$\xi_n = \int \delta_{x,v_n(x)}\, d\mu(x) \quad \text{and} \quad \eta_n = \frac{1}{n} \sum_{j=0}^{n-1} \mathscr{G}_*^j \xi_n.$$

It is clear that $\pi_* \xi_n = \mu$ for every n. It follows that $\pi_* \eta_n = \mu$ for every n, because $\mathscr{G}$ is a skew product over f and the measure μ is f-invariant. By the compactness of $\mathscr{M}(\mu)$, there exists $(n_k)_k \to \infty$ such that $(\eta_{n_k})_k$ converges to some $\eta_I \in \mathscr{M}(\mu)$. Given any $\Psi \in \mathscr{F}$,

$$\begin{aligned} \left| \int (\Psi \circ \mathscr{G})\, d\eta_{n_k} - \int \Psi\, d\eta_{n_k} \right| &= \frac{1}{n_k} \left| \int (\Psi \circ \mathscr{G}^{n_k})\, d\xi_{n_k} - \int \Psi\, d\xi_{n_k} \right| \\ &\leq \frac{2}{n_k} \int \|\Psi(x,\cdot)\|_0\, d\mu(x) = \frac{2}{n_k} \|\Psi\|_1. \end{aligned} \tag{4.8}$$

Observe that $\Psi \circ \mathscr{G} \in \mathscr{F}$, since $\mathscr{G}$ is measurable, $\mathscr{G}_x$ is continuous and

$$\|(\Psi \circ \mathscr{G})(x,\cdot)\|_0 \leq \|\Psi(f(x),\cdot)\|_0$$

for μ-almost every x. Thus, by the definition of the weak* topology, the left-hand side of (4.8) converges to $\left| \int (\Psi \circ \mathscr{G})\, d\eta_I - \int \Psi\, d\eta_I \right|$. The right-hand side converges to zero. So, we have shown that

$$\int (\Psi \circ \mathscr{G})\, d\eta_I = \int \Psi\, d\eta_I \quad \text{for all } \Psi \in \mathscr{F}.$$

This implies that η_I is a $\mathscr{G}$-invariant measure (Exercise 4.3). Finally,

$$\int \Phi\, d\eta_I = \lim_k \int \Phi\, d\eta_{n_k} = \lim_k \int \frac{1}{n_k} \sum_{j=0}^{n_k-1} \Phi(\mathscr{G}^j(x, v_{n_k}(x)))\, d\mu(x)$$
$$= \lim_k \frac{1}{n_k} \int I_{n_k}(x)\, d\mu(x) = \int I\, d\mu$$

(the last step uses the subadditive ergodic theorem). This proves our claim. □

Corollary 4.9 *In the setting of Proposition 4.8, for μ-almost every x there exist $v_I(x) \in P$ and $v_S(x) \in P$ such that*

$$\lim_n \frac{1}{n} \sum_{j=0}^{n-1} \Phi(\mathscr{G}^j(x, v_I(x))) = I(x) \quad \textit{and} \quad \lim_n \frac{1}{n} \sum_{j=0}^{n-1} \Phi(\mathscr{G}^j(x, v_S(x))) = S(x).$$

Proof Up to replacing Φ by $-\Phi$, it suffices to prove either one of the two claims. It is clear that, for every $v \in P$ and μ-almost every x,

$$I(x) \le \liminf_n \frac{1}{n} \sum_{j=0}^{n-1} \Phi(\mathscr{G}^j(x, v)) \le \limsup_n \frac{1}{n} \sum_{j=0}^{n-1} \Phi(\mathscr{G}^j(x, v)) \le S(x). \tag{4.9}$$

By the Birkhoff ergodic theorem, given any $\mathscr{G}$-invariant measure η,

$$\tilde{\Phi}(x, v) = \lim_n \frac{1}{n} \sum_{j=0}^{n-1} \Phi(\mathscr{G}^j(x, v)) \tag{4.10}$$

exists for η-almost every point (x, v), and satisfies $\int \tilde{\Phi}\, d\eta = \int \Phi\, d\eta$. Taking $\eta = \eta_I$, this gives that $\int \tilde{\Phi}\, d\eta = \int I\, d\mu$. In view of (4.9), this implies that the measurable set

$$E = \{(x, v) \in M \times P : \tilde{\Phi}(x, v) = I(x)\}$$

has full η_I-measure. By Proposition 4.5, the set $\pi(E)$ is measurable. Moreover, it has full μ-measure: since $\pi_* \eta_I = \mu$,

$$\mu(\pi(E)) = \eta_I(\pi^{-1}(\pi(E))) \ge \eta_I(E) = 1.$$

Now, it is clear that every $x \in \pi(E)$ satisfies the claim relative to I and v_I. □

Remark 4.10 When $\mathscr{G}$ is invertible, we may replace (4.10) by

$$\tilde{\Phi}(x, v) = \lim_{n \to \pm\infty} \frac{1}{n} \sum_{j=0}^{n-1} \Phi(\mathscr{G}^j(x, v))$$

(the two limits exist and are equal almost everywhere). Then we get

$$\lim_{n\to\pm\infty}\frac{1}{n}\sum_{j=0}^{n-1}\Phi(\mathscr{G}^j(x,v_I(x))) = I(x) \text{ and } \lim_{n\to\pm\infty}\frac{1}{n}\sum_{j=0}^{n-1}\Phi(\mathscr{G}^j(x,v_S(x))) = S(x)$$

in the conclusion of Corollary 4.9.

4.2.4 Applications to linear cocycles

Let us go back to linear cocycles. We are going to deduce:

Proposition 4.11 *Let $x \mapsto V_x$ be a measurable invariant sub-bundle for the cocycle $F : M \times \mathbb{R}^d \to M \times \mathbb{R}^d$. Then, for μ-almost every point x,*

(a) $\lim_n \frac{1}{n}\log\|(A^n(x) \mid V_x)^{-1}\|^{-1} = \min\{\lambda(x,v) : v \in V_x \setminus \{0\}\}$;

(b) $\lim_n \frac{1}{n}\log\|(A^n(x) \mid V_x)\| = \max\{\lambda(x,v) : v \in V_x \setminus \{0\}\}$.

Proof Up to restricting to convenient invariant measurable subsets of M, and considering the normalized restriction of μ to such subsets, we may suppose that the dimension of V_x is constant. Let $l = \dim V_x$ for every x. By Exercise 4.4 and Gram–Schmidt, we may find measurable functions $\{v_1(x),\dots,v_l(x)\}$ that constitute a orthonormal basis of V_x at every point. Using such a basis, we may identify V_x with $\mathbb{R}^l$ through an isometry.

Denote $D(x) = A(x) \mid V_x$ and let $G : M \times \mathbb{R}^l \to M \times \mathbb{R}^l$ be the linear cocycle induced by the map $D : M \to \mathrm{GL}(l)$ over f. Clearly, $\|D(x)\| \le \|A(x)\|$ and $\|D(x)^{-1}\| \le \|A(x)^{-1}\|$ and so the hypotheses imply $\log^+\|D^{\pm 1}\| \in L^1(\mu)$. We also consider the projectivization

$$\mathscr{G} : M \times \mathbb{PR}^l \to M \times \mathbb{PR}^l$$

of G. Consider $\Phi : M \times \mathbb{PR}^l \to \mathbb{R}$, defined by

$$\Phi(x,[v]) = \log\frac{\|D(x)v\|}{\|v\|}.$$

It is clear that $\Phi \in \mathscr{F}$; that is, Φ is measurable and $\Phi(x,\cdot) \in C^0(\mathbb{PR}^d)$ for every x. For any vector $v \in \mathbb{R}^l \setminus \{0\}$ and any $n \ge 0$,

$$\limsup_n \frac{1}{n}\sum_{j=0}^{n-1}\Phi(\mathscr{G}^j(x,[v])) = \limsup_n \frac{1}{n}\log\frac{\|D^n(x)v\|}{\|v\|} = \lambda(x,v).$$

Moreover,

$$I_n(x) = \log\|D^n(x)^{-1}\|^{-1} \quad \text{and} \quad S_n(x) = \log\|D^n(x)\|$$

for every $n \geq 0$, and so

$$I(x) = \lambda_-(x) \quad \text{and} \quad S(x) = \lambda_+(x).$$

According to Lemma 4.4(i), we have $\lambda_-(x) \leq \lambda(x,v) \leq \lambda_+(x)$ for every $v \neq 0$. So, the conclusion of Corollary 4.9 means that

$$\min\{\lambda(x,v) : v \in \mathbb{R}^l \setminus \{0\}\} = \lambda_-(x) \text{ and } \max\{\lambda(x,v) : v \in \mathbb{R}^l \setminus \{0\}\} = \lambda_+(x),$$

as we wanted to prove. □

Remark 4.12 If f is invertible, using Remark 4.10 we get:

(a) $\lim_{n\to\pm\infty} \frac{1}{n} \log \|(A^n(x) \mid V_x)^{-1}\|^{-1} = \min\{\lambda(x,v) : v \in V_x \setminus \{0\}\}$;

(b) $\lim_{n\to\pm\infty} \frac{1}{n} \log \|(A^n(x) \mid V_x)\| = \max\{\lambda(x,v) : v \in V_x \setminus \{0\}\}$.

The special case $V_x = \mathbb{R}^d$ in Proposition 4.11 gives:

Corollary 4.13 $\lambda_+(x) = \lambda_1(x)$ *and* $\lambda_-(x) = \lambda_{k(x)}(x)$ *for* μ*-almost every* x.

4.2.5 Dimension reduction

Let us consider the following situation. Let $x \mapsto V_x$ be a measurable invariant sub-bundle and $\alpha(x) < \beta(x)$ be measurable invariant functions such that

(i) $\lambda(x,v) \leq \alpha(x)$ for every $v \in V_x \setminus \{0\}$;
(ii) $\lambda(x,u) \geq \beta(x)$ for every $u \in \mathbb{R}^d \setminus V_x$;

for μ-almost every x. By Proposition 4.11, condition (i) implies

(iii) $\lim_n \frac{1}{n} \log \|A^n(x) \mid V_x\| \leq \alpha(x)$.

Let $V_x^\perp$ denote the orthogonal complement of V_x. Note that $x \mapsto V_x^\perp$ is measurable, since $x \mapsto V_x$ is taken to be measurable and the *orthogonal complement* map $\perp$: $\mathrm{Gr}(l,d) \to \mathrm{Gr}(d-l,d)$ is a diffeomorphism for every l. Let

$$A(x) = \begin{pmatrix} B(x) & 0 \\ C(x) & D(x) \end{pmatrix} \tag{4.11}$$

be the expression of A relative to the direct sum decomposition $\mathbb{R}^d = V_x^\perp \oplus V_x$. Clearly, $D(x)$ is just the restriction $A(x) \mid V_x$ of $A(x)$ to the invariant sub-bundle. It is also clear that the norms of $B(x)$, $C(x)$ and $D(x)$, and their inverses, are bounded by the norms of $A(x)^{\pm 1}$. Thus, the hypotheses ensure that

$$\log \|B^{\pm 1}\|,\ \log \|C^{\pm 1}\|,\ \log \|D^{\pm 1}\| \in L^1(\mu). \tag{4.12}$$

Proposition 4.14 *For μ-almost every x, any $u \in V_x^\perp \setminus \{0\}$ and any $v \in V_x$,*

(a) $\limsup_n \frac{1}{n} \log \|B^n(x)u\| = \limsup_n \frac{1}{n} \log \|A^n(x)(u+v)\|$.

(b) *if* $\lim_n \frac{1}{n} \log \|B^n(x)u\|$ *exists then* $\lim_n \frac{1}{n} \log \|A^n(x)(u+v)\|$ *exists for all* $v \in V_x$, *and the two limits coincide.*

Proof First, we prove (a). By Exercise 3.6 and the assumptions (i) and (ii),

$$\begin{aligned}\limsup_n \frac{1}{n} \log\|A^n(x)(u+v)\| &\leq \limsup_n \frac{1}{n} \log\left(\|A^n(x)u\| + \|A^n(x)v\|\right) \\ &= \max\left\{\limsup_n \frac{1}{n} \log\|A^n(x)u\|, \limsup_n \frac{1}{n} \log\|A^n(x)v\|\right\} \\ &= \limsup_n \frac{1}{n} \log\|A^n(x)u\|\end{aligned}$$

and, analogously,

$$\begin{aligned}\limsup_n \frac{1}{n} \log\|A^n(x)u\| &\leq \limsup_n \frac{1}{n} \log\left(\|A^n(x)(u+v)\| + \|A^n(x)v\|\right) \\ &= \max\left\{\limsup_n \frac{1}{n} \log\|A^n(x)(u+v)\|, \limsup_n \frac{1}{n} \log\|A^n(x)v\|\right\} \\ &= \limsup_n \frac{1}{n} \log\|A^n(x)(u+v)\|.\end{aligned}$$

This proves that, for any $u \in V_x^\perp \setminus \{0\}$ and $v \in V_x$,

$$\limsup_n \frac{1}{n} \log\|A^n(x)(u+v)\| = \limsup_n \frac{1}{n} \log\|A^n(x)u\|.$$

So, from now on we consider $v = 0$. We will need the following fact:

Lemma 4.15 *Given any $\varepsilon > 0$, there exists a measurable function $d_\varepsilon(x) > 0$ such that*

$$\|D^n(f^m(x))\| \leq d_\varepsilon(x) e^{\alpha(x)n+(m+n)\varepsilon} \quad \textit{for every } m,n \geq 0.$$

Proof Define

$$b_\varepsilon(x) = \sup\{\|D^n(x)\| \, e^{-n(\alpha(x)+\varepsilon)} : n \geq 0\}.$$

Observe that $1 \leq b_\varepsilon(x) < \infty$ at μ-almost every point, by condition (iii). From

$$b_\varepsilon(f(x)) = \sup\{\|D^n(f(x))\| \, e^{-n(\alpha(x)+\varepsilon)} : n \geq 0\} \tag{4.13}$$

we get that

$$b_\varepsilon(f(x)) \geq \|D(x)\|^{-1} e^{\alpha(x)+\varepsilon} \sup\{\|D^n(x)\| e^{-n(\alpha(x)+\varepsilon)} : n \geq 1\}.$$

There are two possibilities. If the supremum on the right-hand side coincides with $b_\varepsilon(x)$, then this yields

$$\log b_\varepsilon(f(x)) \geq \log b_\varepsilon(x) + \log \|D(x)\|^{-1} + \alpha(x) + \varepsilon.$$

Otherwise, if the supremum in the definition of $b_\varepsilon(x)$ is attained at $n = 0$, then $b_\varepsilon(x) = 1 \leq b_\varepsilon(f(x))$. In any event,

$$\log b_\varepsilon(f(x)) - \log b_\varepsilon(x) \geq \min\{-\log^+ \|D(x)\| + \alpha(x) + \varepsilon, 0\}. \tag{4.14}$$

Similarly, (4.13) yields $b_\varepsilon(f(x)) \leq \|D(x)^{-1}\| e^{\alpha(x)+\varepsilon} b_\varepsilon(x)$, and so

$$\log b_\varepsilon(f(x)) - \log b_\varepsilon(x) \leq \log^+ \|D(x)^{-1}\| + \alpha(x) + \varepsilon. \tag{4.15}$$

The inequalities (4.14)–(4.15), together with (4.12), ensure that $\log b_\varepsilon \circ f - \log b_\varepsilon$ is in $L^1(\mu)$. Then, using Corollary 3.11,

$$\lim_m \frac{1}{m} \log b_\varepsilon(f^m(x)) = 0$$

Let $d_\varepsilon(x) = \sup\{b_\varepsilon(f^m(x))e^{-\varepsilon m} : m \geq 0\}$. The previous relation ensures that $d_\varepsilon(x)$ is finite at μ-almost every point. By definition, $b_\varepsilon(f^m(x)) \leq d_\varepsilon(x)e^{\varepsilon m}$ and

$$\|D^n(f^m(x))\| \leq b_\varepsilon(f^m(x))e^{n(\alpha(x)+\varepsilon)} \leq d_\varepsilon(x)e^{\alpha(x)n+(m+n)\varepsilon}$$

for every $m, n \geq 0$. □

Going back to the proof of Proposition 4.14, observe that

$$A^n(x) = \begin{pmatrix} B^n(x) & 0 \\ C_n(x) & D^n(x) \end{pmatrix} \quad \text{for every } n \geq 0,$$

where

$$C_n(x) = \sum_{j=0}^{n-1} D^{n-j-1}(f^{j+1}(x))C(f^j(x))B^j(x). \tag{4.16}$$

Given $x \in M$ and $u \in V_x \setminus \{0\}$, consider

$$\gamma = \max\left\{\alpha(x), \limsup_n \frac{1}{n} \log \|B^n(x)u\|\right\}$$

In particular,

$$\limsup_n \frac{1}{n} \log \|B^n(x)u\| \leq \gamma \tag{4.17}$$

which implies that, given any $\varepsilon > 0$, there exists a real number b_ε such that

$$\|B^j(x)u\| \leq b_\varepsilon e^{j(\gamma+\varepsilon)} \quad \text{for every } j. \tag{4.18}$$

By Corollary 3.11 and (4.12), there exists a measurable function c_ε such that

$$\|C(f^j(x))\| \le c_\varepsilon(x)e^{j\varepsilon} \quad \text{for every } j. \tag{4.19}$$

By Lemma 4.15, there exists a measurable function d_ε such that

$$\|D^i(f^j(x))\| \le d_\varepsilon(x)e^{i\alpha(x)+(i+j)\varepsilon} \quad \text{for every } i \text{ and } j. \tag{4.20}$$

Substituting the estimates (4.18)–(4.20) in (4.16), we find that

$$\|C_n(x)u\| \le \sum_{j=0}^{n-1} d_\varepsilon(x)e^{(n-j-1)\alpha(x)+n\varepsilon}c_\varepsilon(x)e^{j\varepsilon}b_\varepsilon e^{j(\gamma+\varepsilon)} \le na_\varepsilon e^{n(\gamma+3\varepsilon)},$$

where $a_\varepsilon = b_\varepsilon c_\varepsilon(x)d_\varepsilon(x)$. Consequently,

$$\limsup_n \frac{1}{n}\log\|C_n(x)u\| \le \gamma+3\varepsilon. \tag{4.21}$$

Since $A^n(x)u = (B^n(x)u, C_n(x)u)$, we have that

$$\|A^n(x)u\|^2 = \|B^n(x)u\|^2 + \|C_n(x)u\|^2,$$

So, by Exercise 3.6, the inequalities (4.17) and (4.21) give

$$\limsup_n \frac{1}{n}\log\|B^n(x)u\| \le \limsup_n \frac{1}{n}\log\|A^n(x)u\| \le \gamma+3\varepsilon.$$

Since $\varepsilon > 0$ is arbitrary, this implies that

$$\limsup_n \frac{1}{n}\log\|B^n(x)u\| \le \limsup_n \frac{1}{n}\log\|A^n(x)u\| \le \gamma. \tag{4.22}$$

Now, the assumption (ii) yields

$$\alpha(x) < \beta(x) \le \limsup_n \frac{1}{n}\log\|A^n(x)u\| \le \gamma$$

and so $\alpha(x)$ is strictly smaller than γ. So, by the definition of γ,

$$\limsup_n \frac{1}{n}\log\|B^n(x)u\| = \gamma. \tag{4.23}$$

The two relations (4.22) and (4.23) yield part (a) of the proposition.

Now we can readily deduce part (b). Note that

$$\|A^n(x)(u+v)\|^2 = \|B^n(x)u\|^2 + \|C_n(x)u + D^n(x)v\|^2,$$

since $A^n(x)(u+v) = (B^n(x)u, C_n(x)u + D^n(x)v)$. So, by Exercise 3.6,

$$\begin{aligned}\liminf_n \frac{1}{n} \log \|A^n(x)(u+v)\| &\geq \max\{\liminf_n \frac{1}{n} \log \|B^n(x)u\|, \liminf_n \frac{1}{n} \log \|C_n(x)u + D^n(x)v\|\} \\ &\geq \liminf_n \frac{1}{n} \log \|B^n(x)u\| = \limsup_n \frac{1}{n} \log \|B^n(x)u\|.\end{aligned}$$

Then the claim follows immediately from part (a) of the proposition. This finishes the proof of Proposition 4.14. □

4.2.6 Completion of the proof

Keep in mind that our goal, to finish the proof of Theorem 4.1, is to show that

$$\lim_n \frac{1}{n} \log \|A^n(x)v\| \tag{4.24}$$

exists for μ-almost every $x \in M$, every $v \in V_x^i \setminus V_x^{i+1}$ and every $1 \leq i \leq k$. It will follow that the limit is equal to $\lambda(x,v) = \lambda_i(x)$, of course.

Up to replacing M by suitable invariant measurable subsets, and considering the normalized restriction of μ to each of such subsets, we may suppose that $k(x)$ is independent of x, and so is the dimension $l \geq 1$ of the invariant subbundle $V_x = V_x^k$. Let $\alpha(x) = \lambda_k(x)$ and $\beta(x) = \lambda_{k-1}(x)$. It is clear that conditions (i) and (ii) in Section 4.2.5 are satisfied in this context, and so we will be able to use Proposition 4.14. Moreover, Proposition 4.11 implies that, for μ-almost every x,

$$\lim_n \frac{1}{n} \log \|(A^n(x) \mid V_x)^{-1}\|^{-1} = \lambda_k(x) = \lim_n \frac{1}{n} \log \|A^n(x) \mid V_x\|$$

and so,

$$\lim_n \frac{1}{n} \log \|A^n(x)v\| = \lambda_k(x) \quad \text{for all } v \in V_x \setminus \{0\}. \tag{4.25}$$

Consider the expression (4.11) of A relative to the direct sum decomposition $\mathbb{R}^d = V_x^\perp \oplus V_x$. Using a measurable orthonormal basis $\{w_1(x), \ldots, w_{d-l}(x)\}$, we may identify $V_x^\perp$ with $\mathbb{R}^{d-l}$ and view each $B(x)$ as an element of $\mathrm{GL}(d-l)$, depending measurably on the point x. Recall that $\log^\pm \|B\| \in L^1(\mu)$, by (4.12). Define

$$U_x^i = V_x^\perp \cap V_x^i, \quad \text{for every } i = 1, \ldots, k.$$

By Proposition 4.14(a), for every $1 \leq i \leq k-1$ and $u \in U_x^i \setminus U_x^{i+1}$,

$$\limsup_n \frac{1}{n} \log \|B^n(x)u\| = \limsup_n \frac{1}{n} \log \|A^n(x)u\| = \lambda_i(x).$$

Thus, $\mathbb{R}^{d-l} = U_x^1 \supsetneq \cdots \supsetneq U_x^{k-1} \supsetneq \{0\}$ is the Oseledets flag of B, and the Lyapunov exponents are $\lambda_1(x), \dots, \lambda_{k-1}(x)$. By induction on k,

$$\lim_n \frac{1}{n} \log \|B^n(x)u\| = \lambda_i(x) \quad \text{for all } u \in U_x^i \setminus U_x^{i+1}$$

and every $i = 1, \dots, k-1$. By Proposition 4.14(b), it follows that

$$\lim_n \frac{1}{n} \log \|A^n(x)v\| = \lambda_i(x) \quad \text{for all } v \in V_x^i \setminus V_x^{i+1} \tag{4.26}$$

and every $i = 1, \dots, k-1$. The relations (4.25) and (4.26) contain (4.24).

The proof of Theorem 4.1 is complete.

4.3 Proof of the two-sided theorem

Now let us prove Theorem 4.2. We have seen in Remark 4.12 that the limits are not affected when one takes $n \to -\infty$ instead. The main remaining steps for proving Theorem 4.2 are: to upgrade the Oseledets flag to an invariant decomposition (Section 4.3.1) and to prove the subexponential decay of the angles (Section 4.3.2).

4.3.1 Upgrading to a decomposition

We pick-up where we left, in Section 4.2.6. It is no restriction to suppose that $k(x)$ and the dimension $l \geq 1$ of the invariant sub-bundle $V_x = V_x^{k(x)}$ are independent of x. Let $\alpha(x) = \lambda_k(x)$ and $\beta(x) = \lambda_{k-1}(x)$. Consider the expression (4.11) of A relative to the decomposition $\mathbb{R}^d = V_x^\perp \oplus V_x$. By Remark 4.12,

$$\lim_{n \to \pm\infty} \frac{1}{n} \log \|D^n(x)^{-1}\|^{-1} = \lim_{n \to \pm\infty} \frac{1}{n} \log \|D^n(x)\| = \alpha(x) \tag{4.27}$$

for μ-almost every x. Similarly, since $\lambda_1, \dots, \lambda_{k-1}$ are the Lyapunov exponents of B (see Section 4.2.6), Remark 4.12 gives

$$\lim_{n \to \pm\infty} \frac{1}{n} \log \|B^n(x)^{-1}\|^{-1} = \beta(x) \quad \text{for } \mu\text{-almost every } x. \tag{4.28}$$

We are going to use these facts in the proof of the following result.

Proposition 4.16 *If $f : M \to M$ is invertible then there exists a measurable invariant sub-bundle $x \mapsto W_x$ such that $\mathbb{R}^d = W_x \oplus V_x$ for μ-almost every x.*

Proof Let $\mathscr{L}$ be the space of all measurable maps $L : x \mapsto L_x$ assigning to μ-almost every x a linear map $L_x : V_x^\perp \to V_x$. The *graph transform* $T : \mathscr{L} \to \mathscr{L}$ is the transformation characterized by the condition that, for every $x \in M$, *the* *image of the graph of L_x under $A(x)$ coincides with the graph of $T(L)_{f(x)}$.* Using the expression (4.11), this condition translates to the following relation:

$$T(L)_{f(x)} = \big[C(x) + D(x)L_x\big]B(x)^{-1} \quad \text{for every } x. \tag{4.29}$$

Clearly, the graph of some $L \in \mathscr{L}$ is an invariant sub-bundle if and only if $T(L)_x = L_x$ for μ-almost every x. Thus, to prove the proposition it suffices to find such an (essentially) fixed point of the graph transform.

With that in mind, let us rewrite (4.29) as

$$T(L)_y = R_y + S(L)_y \quad \text{for every } y, \tag{4.30}$$

where $R: y \mapsto R_y = C(f^{-1}(y))B(f^{-1}(y))^{-1} = C(f^{-1}(y))B^{-1}(y)$ is an element of $\mathscr{L}$ and $S: \mathscr{L} \to \mathscr{L}$ is the linear operator defined by

$$S(L)_y = D(f^{-1}(y))L_{f^{-1}(y)}B^{-1}(y).$$

We claim that there exist measurable functions $a(x) > 0$ and $\varepsilon(x) > 0$ such that

$$\|S^k(R)_y\| \le a(y)e^{-k\varepsilon(y)} \quad \text{for every } k \ge 0 \text{ and } \mu\text{-almost every } y. \tag{4.31}$$

Assume this fact for a while. Then the sum $L: y \mapsto L_y = \sum_{k=0}^{\infty} S^k(R)_y$ is well defined μ-almost everywhere, and it is a measurable map; in other words, $L \in \mathscr{L}$. Moreover, L is a fixed point of the graph transform:

$$R_y + S(L)_y = R_y + S\Big(\sum_{k=0}^{\infty} S^k(R)\Big)_y = R_y + \sum_{k=1}^{\infty} S^k(R)_y = \sum_{k=0}^{\infty} S^k(R)_y = L_y.$$

We are left to prove the claim (4.31). Begin by observing that, for any $k \ge 0$,

$$S^k(R)_y = D^k(f^{-k}(y))R_{f^{-k}(y)}B^{-k}(y).$$

For each y, take

$$\varepsilon(y) = \frac{1}{5}\big(\beta(y) - \alpha(y)\big).$$

By (4.27) and Lemma 4.15, there exists a measurable function $d(\cdot)$ such that

$$\|D^k(f^{-k}(y))\| \le d(y)e^{k(\alpha(y)+2\varepsilon(y))} \quad \text{for every } k \ge 0. \tag{4.32}$$

Similarly, (4.28) and Lemma 4.15 imply that there exists a measurable function $b(\cdot)$ such that

$$\|B^k(y)^{-1}\| \le b(y)e^{k(-\beta(y)+\varepsilon(y))} \quad \text{for every } k \ge 0. \tag{4.33}$$

The observation (4.12) implies that $\log^+\|R\| \le \log\|C \circ f^{-1}\| + \log^+\|B^{-1}\|$ is in $L^1(\mu)$. Then we may use Corollary 3.11: there exists a measurable function r such that

$$\|R_{f^{-k}(y)}\| \le r(y)e^{k\varepsilon(y)} \quad \text{for every } k \ge 0. \tag{4.34}$$

Let $a(y) = b(y)d(y)r(y)$. Putting (4.32)–(4.34) together,

$$\|S^k(L)_y\| \leq b(y)d(y)r(y)e^{k(\alpha(y)-\beta(y)+4\varepsilon(y))} = a(y)e^{-k\varepsilon(y)}$$

This proves (4.28), and that completes the proof of the proposition. □

The relation (4.27) implies that

$$\lim_{n\to\pm\infty} \frac{1}{n} \log \|A^n(x)v\| = \alpha(x) = \lambda_k(x) \quad \text{for all } v \in V_z \setminus \{0\} \tag{4.35}$$

and μ-almost every x. Identify W_x with $\mathbb{R}^{d-l}$ through some measurable orthonormal basis and let $\tilde{A} : M \to \mathrm{GL}(d-l)$ be given by $\tilde{A}(x) = A(x) \mid W_x$. Define

$$W_x^i = W_x \cap V_x^i \quad \text{for every } i = 1, \ldots, k.$$

By Proposition 4.14(a), for every $1 \leq i \leq k-1$ and $u \in W_x^i \setminus W_x^{i+1}$,

$$\limsup_n \frac{1}{n} \log \|\tilde{A}^n(x)u\| = \limsup_n \frac{1}{n} \log \|A^n(x)u\| = \lambda_i(x).$$

Thus, $W_x = W_x^1 \supsetneq \cdots \supsetneq W_x^{k-1} \supsetneq \{0\}$ is the Oseledets flag of $\tilde{A}$, and the Lyapunov exponents are $\lambda_1(x), \ldots, \lambda_{k-1}(x)$. By induction on k, there exists an $\tilde{A}$-invariant splitting $W_x = E_x^1 \oplus \cdots \oplus E_x^{k-1}$ such that

$$W_x^j = \bigoplus_{i=j}^{k-1} E_x^i \quad \text{for every } j = 1, \ldots, k-1 \tag{4.36}$$

$$\lim_{n\to\pm\infty} \frac{1}{n} \log \|A^n(x)u\| = \lambda_i(x) \quad \text{for all } u \in E_x^i \setminus \{0\} \tag{4.37}$$

and μ-almost every x. Denote $E_x^k = V_x = V_x^k$. Then $\mathbb{R}^d = E_x^1 \oplus \cdots \oplus E_x^{k-1} \oplus E_x^k$ is an A-invariant splitting and (4.36) leads to

$$V_x^j = V_x^k \oplus W_x^j = \bigoplus_{i=j}^{k} E_x^i \quad \text{for every } j = 1, \ldots, k.$$

This proves part (a) of the theorem. Part (b) is contained in (4.35) and (4.37).

4.3.2 Subexponential decay of angles

All that is left to do is to prove part (c) of Theorem 4.2. As explained before, it is no restriction to assume that $k(x)$ and the dimensions of the Oseledets subspaces are constant μ-almost everywhere. Given disjoint subsets I and J of $\{1, \ldots, k\}$, define

$$\phi(x) = \left| \sin \measuredangle \left(\bigoplus_{i\in I} E_x^i, \bigoplus_{j\in J} E_x^j \right) \right|.$$

Since the Oseledets subspaces are invariant,

$$\frac{\phi(f(x))}{\phi(x)} = \frac{|\sin\measuredangle(A(x)\bigoplus_{i\in I}E_x^i, A(x)\bigoplus_{j\in J}E_x^j)|}{|\sin\measuredangle(\bigoplus_{i\in I}E_x^i, \bigoplus_{j\in J}E_x^j)|}.$$

By elementary linear algebra (Exercise 3.5), the right-hand side is bounded above by $\|A(x)\|\|A(x)^{-1}\|$ and below by $(\|A(x)\|\|A(x)^{-1}\|)^{-1}$. In particular, the hypotheses $\log^+\|A^{\pm 1}\| \in L^1(\mu)$ imply that $\log\phi\circ f - \log\phi$ is μ-integrable. Equivalently, $\log\phi\circ f^{-1} - \log\phi$ is μ-integrable. Applying Corollary 3.11, both to f and to f^{-1} we conclude that

$$\lim_{n\to\pm\infty}\frac{1}{n}\phi(f^n(x)) = 0,$$

which is what we wanted to prove. The proof of Theorem 4.2 is complete.

4.3.3 Consequences of subexponential decay

Let us begin by pointing out that we proved a stronger fact than was stated in part (b) of the theorem, namely:

$$\lim_{n\to\pm\infty}\frac{1}{n}\log\|(A^n(x)\mid E_x^i)^{-1}\|^{-1} = \lim_{n\to\pm\infty}\frac{1}{n}\log\|A^n(x)\mid E_x^i\| = \lambda_i(x) \tag{4.38}$$

for μ-almost every x (this implies, for instance, that the limit in (b) is uniform over all unit vectors). It follows (Exercise 4.7) that

$$\lim_{n\to\pm\infty}\frac{1}{n}\log\left|\det(A^n(x)\mid E_x^i)\right| = \lambda_i(x)\dim E_x^i$$

for μ-almost every x. Moreover, using Theorem 4.2(c) and Exercise 4.7,

$$\lim_{n\to\pm\infty}\frac{1}{n}\log\left|\det\left(A^n(x)\mid\bigoplus_{i\in I}E_x^i\right)\right| = \sum_{i\in I}\lambda_i(x)\dim E_x^i \tag{4.39}$$

for any $I\subset\{1,\dots,k(x)\}$. For example,

$$\lim_{n\to\pm\infty}\frac{1}{n}\log|\det A^n(x)| = \sum_{i=1}^{k(x)}\lambda_i(x)\dim E_x^i \tag{4.40}$$

for μ-almost every $x\in M$. In particular, if $|\det A|\equiv 1$ then the sum of all Lyapunov exponents, counted with multiplicity, is identically zero. Analogously,

$$\begin{aligned}\lim_{n\to\pm\infty}\frac{1}{n}\log\left|\det(A^n(x)\mid E_x^u)\right| &= \sum_{\lambda_i(x)>0}\lambda_i(x)\dim E_x^i\\ \lim_{n\to\pm\infty}\frac{1}{n}\log\left|\det(A^n(x)\mid E_x^s)\right|, &= \sum_{\lambda_i(x)<0}\lambda_i(x)\dim E_x^i\end{aligned} \tag{4.41}$$

where $E_x^u = \bigoplus_{\lambda_i(x)>0} E_x^i$ is the *unstable bundle* and $E_x^s = \bigoplus_{\lambda_i(x)<0} E_x^i$ is the *stable bundle*.

One can interpret and expand these conclusions, using the formalism of exterior linear algebra. Let us recall a few basic notions involved, referring the reader to Bourbaki [39, § III.5] for more information.

Let $E = \mathbb{R}^d$ and $0 \le l \le d$. The *exterior l-power* $\Lambda^l E$ of E is the vector space of alternating l-linear forms $\omega : E^* \times \cdots \times E^* \to \mathbb{R}$ on the dual space E^* (a 0-linear form is just a real constant). Its elements are called *l-vectors*. Define the *exterior product* of vectors $v_1, \ldots, v_l \in E$ to be the alternating l-linear form $v_1 \wedge \cdots \wedge v_l : E^* \times \cdots \times E^* \to \mathbb{R}$ defined by

$$(v_1 \wedge \cdots \wedge v_l)(\phi_1, \ldots, \phi_l) = \sum_{\sigma} \operatorname{sgn}(\sigma)\, \phi_{\sigma(1)}(v_1) \cdots \phi_{\sigma(l)}(v_l),$$

where the sum is over all permutations of $\{1, \ldots, l\}$. These are special elements of $\Lambda^l E$, that we call *decomposable l-vectors*. Every l-vector is a sum of decomposable l-vectors. Moreover, if $\{e_j : j = 1, \ldots, d\}$ is a basis of E then $\{e_{j_1} \wedge \cdots \wedge e_{j_l} : 1 \le j_1 < \cdots < j_l \le d\}$ is a basis of $\Lambda^l E$. Thus,

$$\dim \Lambda^l E = \binom{d}{l}.$$

Every linear map $L : E \to E$ induces a linear map $\Lambda^l L : \Lambda^l E \to \Lambda^l E$, by

$$\Lambda^l L(\omega) : (\phi_1, \ldots, \phi_l) \mapsto \omega(\phi_1 \circ L, \ldots, \phi_l \circ L), \tag{4.42}$$

for $\omega \in \Lambda^l E$ and $\phi_1, \ldots, \phi_l \in E^*$. This map preserves the subset $\Lambda^l_{\mathrm{d}} E$ of decomposable l-vectors, since

$$\Lambda^l L(v_1 \wedge \cdots \wedge v_l) = L(v_1) \wedge \cdots \wedge L(v_l).$$

Recall that $\mathrm{Gr}(l,d)$ denotes the Grassmannian manifold, whose elements are the l-dimensional subspaces of E. Since

$$v_1 \wedge \cdots \wedge v_l \neq 0 \quad \Leftrightarrow \quad \{v_1, \ldots, v_l\} \text{ is linearly independent},$$

there is a well-defined map $\Psi : \Lambda^l_{\mathrm{d}} E \setminus \{0\} \to \mathrm{Gr}(l,d)$ assigning to each non-zero decomposable l-vector $v_1 \wedge \cdots \wedge v_l$ the linear subspace spanned by the vectors $v_1, \ldots, v_l$. Two l-vectors have the same image under Ψ if and only if one is a multiple of the other. Hence, this map factors through a bijection

$$\mathbb{P}\Lambda^l_{\mathrm{d}} E \to \mathrm{Gr}(l,d),$$

that we can use to identify the projectivization of the space of decomposable l-vectors with the Grassmannian manifold. Then (4.42) corresponds to the natural action $F \mapsto L(F)$ of L on the Grassmannian.

Now take E to be endowed with some inner product $(v,w) \mapsto v \cdot w$. For decomposable l-vectors, define

$$(v_1 \wedge \cdots \wedge v_l) \cdot (w_1 \wedge \cdots \wedge w_l) = \det\left(v_i \cdot w_j\right)_{1 \le i,j \le d}. \tag{4.43}$$

The right-hand side does not depend on the particular decomposition of the l-vectors as exterior products. Moreover, (4.43) extends to an inner product on the exterior power $\Lambda^l E$. If $\{e_j : j = 1,\ldots,d\}$ is an orthonormal basis of E then $\{e_{j_1} \wedge \cdots \wedge e_{j_l} : 1 \le j_1 < \cdots < j_l \le d\}$ is an orthonormal basis of $\Lambda^l E$. The norm $\|v_1 \wedge \cdots \wedge v_l\|$ of a non-zero decomposable l-vector coincides with its *volume*, defined as the unique positive real number $\operatorname{vol}(v_1,\ldots,v_l)$ such that

$$v_1 \wedge \cdots \wedge v_l = \pm \operatorname{vol}(v_1,\ldots,v_l)\, u_1 \wedge \cdots \wedge u_l,$$

for any orthonormal basis $\{u_1,\ldots,u_l\}$ of the subspace $V = \Psi(v_1 \wedge \cdots \wedge v_l)$. This definition does not depend on the choice of this basis. The *determinant* of a linear isomorphism $L : E \to E$ restricted to V is given, in absolute value, by

$$\left|\det(L \mid V)\right| = \frac{\operatorname{vol}(L(v_1),\ldots,L(v_l))}{\operatorname{vol}(v_1,\ldots,v_l)} = \frac{\|\Lambda^l L(v_1 \wedge \cdots \wedge v_l)\|}{\|v_1 \wedge \cdots \wedge v_l\|}. \tag{4.44}$$

Proposition 4.17 *Let $F : M \times \mathbb{R}^d \to M \times \mathbb{R}^d$, $F(x,v) = (f(x),A(x)v)$ be an invertible linear cocycle and let $\chi_1(x) \le \cdots \le \chi_d(x)$ be the Lyapunov exponents of F, counted with multiplicity. For each $0 \le l \le d$, let $\Lambda^l F$ be the linear cocycle induced by F on the exterior l-power; that is,*

$$\Lambda^l F : M \times \Lambda^l(\mathbb{R}^d) \to M \times \Lambda^l(\mathbb{R}^d), \quad \Lambda^l F(x,\omega) = (f(x),\Lambda^l A(x)\omega).$$

The Lyapunov exponents of the linear cocycle $\Lambda^l F$, counted with multiplicity, are the sums $\chi_{i_1}(x) + \cdots + \chi_{i_l}(x)$, with $1 \le i_1 < \cdots < i_l \le d$.

The proof of the proposition goes as follows. Let $\lambda_1(x) > \cdots > \lambda_k(x)$ be the Lyapunov exponents and $\mathbb{R}^d = E_x^1 \oplus \cdots \oplus E_x^k$ be the Oseledets decomposition of F. We may suppose that k and the dimensions $d_m = \dim E_x^m$ are constant μ-almost everywhere. For each $(l_1,\ldots,l_k)$ with $0 \le l_m \le d_m$ and $\sum_{m=1}^k l_m = l$, define $\mathscr{E}_x^{l_1,\ldots,l_k}$ to be the subspace of $\Lambda^l(\mathbb{R}^d)$ generated by the l-vectors $v_1 \wedge \cdots \wedge v_l$ with

$$v_j \in E_x^m \quad \text{for} \quad \sum_{i=1}^{m-1} l_i < j \le \sum_{i=1}^{m} l_i. \tag{4.45}$$

Every $x \mapsto \mathscr{E}_x^{l_1,\ldots,l_k}$ is a $\Lambda^l F$-invariant sub-bundle. Moreover, these sub-bundles are in general position with respect to each other, and

$$\dim \mathscr{E}_x^{l_1,\ldots,l_k} = \binom{d_1}{l_1} \cdots \binom{d_k}{l_k}.$$

Then (Exercise 4.5), we have $\bigoplus_{l_1,\dots,l_k} \mathscr{E}_x^{l_1,\dots,l_k} = \Lambda^l(\mathbb{R}^d)$. Take $v_1 \wedge \cdots \wedge v_l$ as in (4.45) and let V_x be the subspace generated by $v_1, \dots, v_l$. Using (4.44),

$$\lim_{n\to\pm\infty} \frac{1}{n} \log \|\Lambda^l A(x)^n (v_1 \wedge \cdots \wedge v_l)\| = \lim_{n\to\pm\infty} \frac{1}{n} \log \left|\det(A^n(x) \mid V_x)\right|.$$

For each $m = 1, \dots, k$, take V_x^m to be the subspace generated by the vectors v_j with $\sum_{i=1}^{m-1} l_i < j \le \sum_{i=1}^{m} l_i$. Then $V_x = \bigoplus_{m=1}^{k} V_x^m$ and, using Theorem 4.2(c) and Exercise 4.7,

$$\lim_{n\to\pm\infty} \frac{1}{n} \log \left|\det(A^n(x) \mid V_x)\right| = \sum_{m=1}^{k} \lim_{n\to\pm\infty} \frac{1}{n} \log \left|\det(A^n(x) \mid V_x^m)\right|.$$

By (4.38) and Exercise 4.7,

$$\lim_{n\to\pm\infty} \frac{1}{n} \log \left|\det(A^n(x) \mid V_x^m)\right| = l_m \lambda_m(x).$$

Putting these three relations together,

$$\lim_{n\to\pm\infty} \frac{1}{n} \log \|\Lambda^l A(x)^n (v_1 \wedge \cdots \wedge v_l)\| = \sum_{m=1}^{k} l_m \lambda_m(x).$$

Since these decomposable k-vectors $v_1 \wedge \cdots \wedge v_l$ generate the space $\mathscr{E}^{l_1,\dots,l_k}$, it follows that the latter is contained in some Oseledets subspace, with Lyapunov exponent $\sum_{m=1}^{k} l_m \lambda_m(x)$.

Thus, the Lyapunov exponents of $\Lambda^l F$ are the numbers $\sum_{m=1}^{k} l_m \lambda_m(x)$, with $0 \le l_m \le d_m$ and $\sum_{m=1}^{k} l_m = l$, each with multiplicity equal to $\dim \mathscr{E}_x^{l_1,\dots,l_k}$ (some of these numbers may coincide, in which case the multiplicities add up). This is a rephrasing of the conclusion of Proposition 4.17.

4.4 Two useful constructions

Here we present a couple of constructions concerning Lyapunov exponents that will be useful later. The reader may choose to skip this section at first reading, and come back to it when these ideas are called for.

4.4.1 Inducing and Lyapunov exponents

Let $Z \subset M$ be a positive μ-measure subset and $g : Z \to Z$ be the first return map, defined by

$$g(x) = f^{r(x)}(x), \quad r(x) = \inf\{n \ge 1 : f^n(x) \in Z\} \tag{4.46}$$

($r(x)$ is finite for μ-almost every x, by Poincaré recurrence). Let ν be the normalized restriction of μ to Z; that is,

$$\nu(E) = \frac{\mu(E)}{\mu(Z)} \quad \text{for every measurable set } E \subset Z. \tag{4.47}$$

Let $G : Z \times \mathbb{R}^d \to Z \times \mathbb{R}^d$ be the linear cocycle over g defined by the function $B : Z \to \mathrm{GL}(d)$, $B(x) = A^{r(x)}(x)$. We are going to see that the Lyapunov exponents of G at each point x are obtained from the Lyapunov exponents of F by multiplication by a constant $c(x) \geq 1$.

Proposition 4.18 *Take the transformation $f : M \to M$ to be invertible.*

(1) *The probability measure ν is g-invariant and we have* $\log^+ \|B^{\pm 1}\| \in L^1(\nu)$ *whenever* $\log^+ \|A^{\pm}\| \in L^1(\mu)$.
(2) *The Oseledets decomposition of G coincides with the restriction of the Oseledets decomposition of F.*
(3) *For ν-almost every $x \in Z$ there exists $c(x) \geq 1$ such that the Lyapunov exponents satisfy $\lambda_j(G,x) = c(x)\lambda_j(F,x)$ for every j.*

Proof For each $j \geq 1$, let Z_j be the subset of points $x \in Z$ such that $r(x) = j$. Then $\{Z_j : j \geq 1\}$ and $\{f^j(Z_j) : j \geq 1\}$ are partitions of full measure subsets of Z. Note also that $g \mid Z_j = f^j \mid Z_j$ for all $j \geq 1$. For any measurable set $E \subset Z$ and any $j \geq 1$,

$$\mu\big(g^{-1}(E \cap f^j(Z_j))\big) = \mu\big(f^{-j}(E \cap f^j(Z_j))\big) = \mu\big(E \cap Z_j\big),$$

because μ is invariant under f. It follows that

$$\mu\big(g^{-1}(E)\big) = \sum_{j=1}^{\infty} \mu\big(g^{-1}(E \cap f^j(Z_j))\big) = \sum_{j=1}^{\infty} \mu\big(E \cap Z_j\big) = \mu(E).$$

So, the normalized restriction ν is invariant under g. Next, from the definition $B(x) = A^{r(x)}(x)$, we get

$$\int_Z \log^+ \|B\| \, d\mu = \sum_{j=1}^{\infty} \int_{Z_j} \log^+ \|A^j\| \, d\mu \leq \sum_{j=1}^{\infty} \sum_{i=0}^{j-1} \int_{Z_j} \log^+ \|A \circ f^i\| \, d\mu.$$

Since μ is invariant under f and the domains $f^i(Z_j)$ are pairwise disjoint for all $0 \leq i \leq j-1$, it follows that

$$\int_Z \log^+ \|B\| \, d\mu \leq \sum_{j=1}^{\infty} \sum_{i=0}^{j-1} \int_{f^i(Z_j)} \log^+ \|A\| \, d\mu \leq \int \log^+ \|A\| \, d\mu.$$

The corresponding bound for the norm of the inverse is obtained in the same way. This proves that $\log^+ \|B^{\pm 1}\|$ is ν-integrable. It is clear that the restriction of the Oseledets decomposition of F is invariant under G. Define

$$c(x) = \lim_{k\to\infty} \frac{1}{k} \sum_{j=0}^{k-1} r(g^j(x)).$$

Note that r is integrable relative to ν:

$$\int_Z r\,d\mu = \sum_{j=1}^{\infty} j\mu(Z_j) = \sum_{j=1}^{\infty}\sum_{i=0}^{j-1} \mu(f^i(Z_j)) \le 1.$$

Thus, by the ergodic theorem, $c(x)$ is well defined at ν-almost every x. It is clear from the definition that $c(x) \ge 1$. Now, given any non-zero vector v and a generic point $x \in Z$,

$$\lim_{k\to\pm\infty} \frac{1}{k} \log \|B^k(x)v\| = c(x) \lim_{n\to\pm\infty} \frac{1}{n} \log \|A^n(x)v\|.$$

(Theorem 4.2 ensures that the limit on the right-hand side exists.) This implies that the Oseledets decomposition of G coincides with the restriction of the Oseledets decomposition of F, and that the Lyapunov exponents of the two cocycles are related by the factor $c(x)$, as claimed in the proposition. □

4.4.2 Invariant cones

The *cone* of radius $a > 0$ around a subspace V of $\mathbb{R}^d$ is defined by

$$C(V,a) = \{v_1 + v_2 \in V \oplus V^\perp : \|v_2\| < a\|v_1\|\}. \tag{4.48}$$

Proposition 4.19 *Assume there exist $\delta > 0$, $b > a > 0$, and an F-invariant measurable decomposition $\mathbb{R}^d = V_x \oplus W_x$, defined at μ-almost every point, such that the dimensions of V_x and W_x are constant,*

$$|\sin\measuredangle(V_x, W_x)| \ge \delta \quad \textit{and} \quad A(x)(C(V_x,b)) \subset C(V_{f(x)},a)$$

for μ-almost every $x \in M$. Then the Lyapunov exponents of F along the sub-bundle V_x are strictly larger than the Lyapunov exponents of F along W_x.

Proof It is no restriction to suppose that the subspace V_x is constant: pick any measurable orthonormal basis $\{v_1(x),\cdots,v_d(x)\}$ of the space $\mathbb{R}^d$ such that $\{v_1(x),\dots,v_l(x)\}$ is a basis of V_x and use it to identify $\mathbb{R}^d$ to itself; then V_x is identified to $\mathbb{R}^l \times \{0\}$. So, from now on we suppose $V_x = V$ for every $x \in M$.

The cone $C(V,b)$ is the union of all the graphs of linear maps $u : V \to V^\perp$ such that $\|u\| < b$. In the space $\mathscr{H}(V,b)$ of such maps, consider the Hilbert

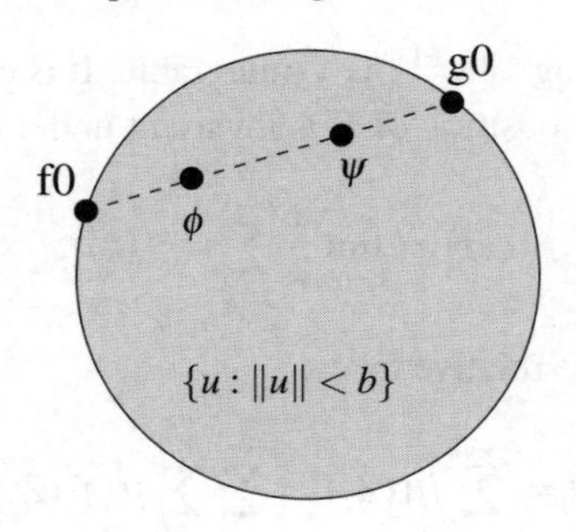

Figure 4.1 Cross-ratio and projective metric

metric (see Birkhoff [26] and [114, Section 12.3.1]), defined by the logarithm of the cross-ratio

$$d(\phi, \psi) = \log \frac{\|\phi - \psi_0\| \|\psi - \phi_0\|}{\|\phi - \phi_0\| \|\psi - \psi_0\|}, \tag{4.49}$$

where ϕ_0 and ψ_0 are the points where the line through ϕ and ψ hits the boundary $\{u : \|u\| = b\}$, denoted in such a way that ϕ_0 is closer to ϕ and ψ_0 is closer to ψ (see Figure 4.1). The hypothesis implies that each $A(x)(\text{graph}(\phi))$, $\phi \in \mathscr{H}(V,b)$ may be written as $\text{graph}(\psi)$ for some $\psi \in \mathscr{H}(V,a) \subset \mathscr{H}(V,b)$. Since $\mathscr{H}(V,a)$ is relatively compact in $\mathscr{H}(V,b)$, it has finite diameter for the Hilbert metric. By a crucial property of the Hilbert metric (see [114, Proposition 12.3.6]), the action $\mathscr{H}(V,b) \to \mathscr{H}(V,b)$ of $A(x)$ thus defined is a θ-contraction with respect to this metric, for some $\theta < 1$ that depends only on an upper bound for the diameter and, hence, can be expressed in terms of a and b alone. It follows (Exercise 4.11) that the width of the iterates $A^n(f^{-n}(x))(C(V,b))$ decays as $Ke^{-\kappa n}$, for some $K > 0$ and $\kappa > 0$ that depend only on a and b. Of course, the intersection of the iterate satisfies

$$\bigcap_{n=1}^{\infty} A^n(f^{-n}(x))(C(V,b)) = V. \tag{4.50}$$

We are going to deduce that any unit vector $v \in V$ is expanded at a definitely faster rate than any unit vector $w \in W_x$. Indeed, one can find $c_0 > 0$ depending only on b and such that $v + c_0 w \in C(V,b)$. Then,

$$A^n(x)v + c_0 A^n(x)w \in C(V, Ke^{-\kappa n}) \quad \text{for every } n \geq 1.$$

Since $A^n(x)w \in W_{f^n(x)}$ and $|\sin\measuredangle(W_{f^n(x)}, V_{f^n(x)})| \geq \delta$, the previous relation implies that

$$\|A^n(x)w\| \leq K_0 e^{-\kappa n} \|A^n(x)v\| \quad \text{for every } n \geq 1,$$

where $K_0 > 0$ depends only on K, c_0, and δ, and so is completely determined by a, b, and δ. It follows that

$$\lim_n \frac{1}{n} \log \|A^n(x)w\| \leq \lim_n \frac{1}{n} \log \|A^n(x)v\| - \kappa$$

for every $v \in V$ and $w \in W_x$, and μ-almost every x. This proves that any Lyapunov exponent of a vector in W_x is smaller, by a definite amount $-\kappa$, than any Lyapunov exponent of a vector in V. □

4.5 Notes

Theorems 4.1 and 4.2 are part of the main result in Oseledets [92]: the results of Oseledets are stated in the slightly broader context of linear cocycles on finite-dimensional vector bundles. Other proofs were given by Raghunathan [101], Ruelle [105] and Ledrappier [80], among others. While the early proofs relied mostly on linear algebra calculations, dynamical proofs were later found by Mañé [89] and Walters [117]. Our presentation is closest to [117] and to its extension to the invertible case by Bochi [28].

The assumption $\log^+ \|A^{\pm 1}\| \in L^1(\mu)$ is unnecessarily strong for Theorem 4.1: it suffices to assume that the function $\log^+ \|A\|$ is integrable; however, in this case the smallest Lyapunov exponent may be $-\infty$. The multiplicative ergodic theorem also extends to some infinite-dimensional cocycles with finitely many non-negative Lyapunov exponents: see Ruelle [106] and Mañé [87].

Proposition 4.17 shows that the subadditive ergodic theorem suffices to identify all the Lyapunov exponents, including their multiplicities, not just the extremal ones: for every $1 \leq l \leq d$, the l-largest Lyapunov exponent (counted with multiplicity) coincides with $\lambda_+(\Lambda^l F) - \lambda_+(\Lambda^{l-1} F)$. This observation was part of the early proofs of the Oseledets theorem.

Most of what follows in this book turns around the multiplicative ergodic theorem. Chapter 6 shows that the Lyapunov exponents may be expressed, in an explicit way, in terms of the invariant measures and the stationary measures of the projective cocycle $\mathbb{P}F$. The invariance principle in Chapter 7 is a general tool for analysing the special situation when all the Lyapunov exponents are equal (and to prove that this is seldom the case). In Chapter 8 we will investigate conditions under which all the Lyapunov exponents are distinct. Finally, in Chapters 9 and 10 we will investigate how Lyapunov exponents vary with the linear cocycle and with the invariant measure μ.

The constructions in Sections 4.4.1 and 4.4.2 will be used in the proof of Propositions 9.13 and 8.3, respectively.

4.6 Exercises

Exercise 4.1 Prove that a map $x \to V_x$ with values in $\mathrm{Gr}(d)$ is measurable if and only if $M_l = \{x \in M : \dim V_x = l\}$ is a measurable set for every $0 \le l \le d$ and, for each l, there exist measurable vector fields $v_i : M_l \to \mathbb{R}^d$, $i = 1, \dots, l$ such that $\{v_1(x), \dots, v_l(x)\}$ is a basis of V_x for every $x \in M_l$.

Exercise 4.2 Prove that (4.5) defines a complete separable norm in $\mathscr{F}$:

(1) Check that $\|\cdot\|_1$ is indeed a norm in $\mathscr{F}$.
(2) Show that any Cauchy sequence in $\mathscr{F}$ admits some subsequence $(\Psi_n)_n$ such that $\|\Psi_n - \Psi_{n+1}\|_1 \le 2^{-2n}$ for every n. Deduce that there exists $N \subset M$ with $\mu(N) = 1$ such that $\Psi(x,v) = \lim_n \Psi_n(x,v)$ exists for every $(x,v) \in N \times P$. Argue that $\psi \in \mathscr{F}$ and use bounded convergence to get that $\|\Psi_n - \Psi\|_1 \to 0$.
(3) Use the fact that $C^0(P)$ is separable, and the assumption that the probability space $(M, \mathscr{B}, \mu)$ is separable, to conclude that $(\mathscr{F}, \|\cdot\|_1)$ is separable (compare Theorem 2 in [120, page 137]).

Exercise 4.3 Show that two probability measures $\eta, \zeta \in \mathscr{M}(\mu)$ coincide if and only if $\int \Psi \, d\eta = \int \Psi \, d\zeta$ for every $\Psi \in \mathscr{F}$.

Exercise 4.4 Prove that the weak* topology on $\mathscr{M}(\mu)$ is metrizable and compact:

(1) Fix a countable dense subset $\{\Psi_k : k \ge 1\}$ of the unit ball of $\mathscr{F}$ and define

$$d(\xi, \eta) = \sum_{k=1}^{\infty} 2^{-k} \left| \int \Psi_k \, d\xi - \int \Phi_k \, d\eta \right| \quad \text{for any } \xi, \eta \in \mathscr{M}(\mu).$$

Show that d is a distance compatible with the weak*-topology.
(2) Given any sequence in $\mathscr{M}(\mu)$, use a diagonal argument to find a subsequence $(\eta_n)_n$ such that $\lim_n \int \Psi_k \, d\eta_n$ exists for every k. Conclude that there exists a unique bounded linear operator $g : \mathscr{F} \to \mathbb{R}$ such that

$$g(\Psi_k) = \lim_n \int \Psi_k \, d\eta_n \quad \text{for every } k.$$

Then $\int \Psi \, d\eta = g(\Psi)$ defines a probability measure η on $M \times P$ that projects down to μ. Moreover, $(\eta_n)_n$ converges to η in the weak* topology.

Exercise 4.5 Check that if the $d_1 + \cdots + d_k = d$ then

$$\binom{d}{l} = \sum_{l_1, \dots, l_k} \binom{d_1}{l_1} \cdots \binom{d_k}{l_k},$$

where the sum is over all $0 \le l_i \le d_i$ with $\sum_{i=1}^{k} l_i = l$.

Exercise 4.6 Show that if $\{v_1,\dots,v_k\}$ and $\{w_1,\dots,w_k\}$ are orthonormal bases for the same subspace V then the matrix $(v_i \cdot w_j)_{i,j}$ is orthonormal.

Exercise 4.7 Show that if $L:\mathbb{R}^d \to \mathbb{R}^d$ is a linear isomorphism and V_1, V_2 are linear subspaces of $\mathbb{R}^d$ then

(1) $\dfrac{1}{\|L\|\,\|L^{-1}\|} \le \dfrac{|\sin\measuredangle(L(V_1),L(V_2))|}{|\sin\measuredangle(V_1,V_2)|} \le \|L\|\,\|L^{-1}\|$;

(2) $\|(L \mid V_1)^{-1}\|^{-d_1} \le |\det(L \mid V_1)| \le \|(L \mid V_1)\|^{d_1}$, where $d_1 = \dim V_1$;

(3) $|\sin\measuredangle(V_1,V_2)| = \|\pi^{-1}\|^{-1}$, where $\pi: V_1 \to V_2$ is the orthogonal projection;

(4) $|\sin\measuredangle(L(V_1),L(V_2))|^{d_1} \le \dfrac{|\det(L \mid V_1 \oplus V_2)|}{|\det(L \mid V_1)|\,|\det(L \mid V_2)|} \le |\sin\measuredangle(V_1,V_2)|^{-d_1}$.

Exercise 4.8 Prove that if μ is ergodic for f then ν is ergodic for the induced map g, and $c(x) = 1/\mu(Z)$ for ν-almost every x.

Exercise 4.9 Extend Proposition 4.18 and Exercise 4.8 to the non-invertible case, using the concept of natural extension (see [114, Section 2.4.2]).

Exercise 4.10 Let V, W be subspaces of $\mathbb{R}^d$ with $\dim V = \dim W$. Prove that:

(1) given $a > 0$ there is $b > 0$ such that $C(W,b) \subset C(V,2a)$ if $W \subset C(V,a)$;
(2) if $b < 1/10$ then $W \subset C(V,b)$ implies $C(V,b) \subset C(W,3b)$.

Exercise 4.11 Let $\rho > 0$ be small. Show that if $\mathcal{H} \subset \mathcal{H}(V,b)$ is contained in the ρ-neighborhood of $\phi \equiv 0$, relative to the Hilbert metric d, then the union C of all graph(ϕ), $\phi \in \mathcal{H}$ is contained in $C(V,\rho/b)$. More generally, relate the d-diameter of a subset $\mathcal{H}$ of $\mathcal{H}(V,b)$ to the width of the corresponding cone C.

Exercise 4.12 Check that one may replace the assumption $|\sin\measuredangle(V_x,W_x)| \ge \delta$ in Proposition 4.19 by a condition of subexponential decay of the angles:

$$\lim_n \frac{1}{n} \log|\sin\measuredangle(A^n(x)V_x, A^n(x)W_x)| = 0.$$

Exercise 4.13 Let $B \in \mathrm{GL}(d)$ be such that $B(C(V,b)) \subset C(V,a)$ for some subspace $V \subset \mathbb{R}^d$ and some $0 < a < b$. Prove that:

(1) B admits a *dominated decomposition* $\mathbb{R}^d = U \oplus S$ with $\dim U = \dim V$; that is, $B(U) = U$, $B(S) = S$, and every eigenvalue of B restricted to U is larger, in norm, than every eigenvalue of B restricted to S;

(2) there is $r > 0$ such that $\tilde{B} = B_1 \cdots B_\ell$ admits a dominated decomposition $\mathbb{R}^d = U_{\tilde{B}} \oplus S_{\tilde{B}}$, for any $\ell \geq 1$ and any $B_1, \ldots, B_\ell$ in the r-neighborhood of B; moreover, $U_{\tilde{B}}$ is uniformly close to U if r is small.

5
Stationary measures

Further development of the theory requires that we exploit in more depth the connection between the Lyapunov exponents of a linear cocycle and the invariant measures of the corresponding projective cocycle, of which we had a brief glimpse in the proof of the multiplicative ergodic theorem. In this chapter we introduce a general formalism that will be very useful towards that end.

Linearity is not relevant at this stage, so we formulate the results for a class of systems more general than linear and projective cocycles, that we call random transformations. The definition and fundamental properties of such systems are discussed in Section 5.1.

In Section 5.2 we introduce the key notion of *stationary measure* for a random transformation. We will see in the next chapter that the measures stationary under the projective cocycle completely determine the Lyapunov exponents of the linear cocycle. The properties of stationary measures of general (possibly non-invertible) random transformations are studied in Sections 5.2 and 5.3.

The invertible case is treated in more detail in Section 5.4 and leads to the important concepts of *u-state* and *s-state*, which are invariant probability measures whose disintegrations along the fibers have special invariance properties. These disintegrations are revisited, from a different angle, in Section 5.5.

5.1 Random transformations

Denote by M the space of sequences $X^{\mathbb{N}}$ (or $X^{\mathbb{Z}}$) in some probability space $(X, \mathscr{X}, p)$, endowed with the product σ-algebra $\mathscr{A} = \mathscr{X}^{\mathbb{N}}$ (or $\mathscr{X}^{\mathbb{Z}}$) and the product measure $\mu = p^{\mathbb{N}}$ (or $\mu = p^{\mathbb{Z}}$). Moreover, let $f : M \to M$ be the shift map on M.

Let $(N, \mathscr{B})$ be a measurable space and take $M \times N$ to be endowed with the product σ-algebra $\mathscr{A} \otimes \mathscr{B}$. Given a set $E \subset M \times N$ and points $x \in M$ and $v \in N$, we will consider the *slices* defined by

$$E_v = \{x \in M : (x,v) \in E\} \quad \text{and} \quad E^x = \{v \in N : (x,v) \in E\}.$$

If $E \subset M \times N$ is measurable then so are $E_v \subset M$ and $E^x \subset N$ (Exercise 5.1).

A *random transformation* (or *locally constant skew product*) over f is a measurable transformation of the form

$$F : M \times N \to M \times N, \quad F(x,v) = (f(x), F_x(v))$$

where F_x depends only on the zeroth coordinate of $x \in M$.

Example 5.1 (Random matrices) Let $X = \{A_1, \dots, A_m\} \subset GL(d)$. Consider the probability measure defined on X by $p = p_1\delta_{A_1} + \cdots + p_m\delta_{A_m}$, with p_1, $\dots$, $p_m > 0$ and $p_1 + \cdots + p_m = 1$. Let $N = \mathbb{R}^d$ and $F_x = x_0$ for every x in $M = X^{\mathbb{Z}}$. Two random transformations associated with this data are relevant for our purposes: the locally constant linear cocycle $F : M \times \mathbb{R}^d \to M \times \mathbb{R}^d$,

$$((\alpha_n)_n, v) \mapsto ((\alpha_{n+1})_n, \alpha_0(v)),$$

and the corresponding projective cocycle $\mathbb{P}F : M \times \mathbb{P}\mathbb{R}^d \to M \times \mathbb{P}\mathbb{R}^d$,

$$((\alpha_n)_n, [v]) \mapsto ((\alpha_{n+1})_n, [\alpha_0(v)]).$$

Example 5.2 Let $F_1, F_2 : S^1 \to S^1$ be homeomorphisms of the circle $N = S^1$. Consider $X = \{1,2\}$ and $M = X^{\mathbb{N}}$. Let $f : M \to M$ be the shift map and $\mu = p^{\mathbb{N}}$ with $p = p_1\delta_1 + p_2\delta_2$. Let $F : M \times N \to M \times N$, $F(x,v) = (f(x), F_{x_0}(v))$. Then $F^n(x,v) = (f^n(x), F_{i(n-1)} \cdots F_{i(0)}(v))$ where the $F_{i(j)} \in \{F_1, F_2\}$ are independent random variables with identical distribution p.

Let $F : M \times N \to M \times N$ be a random transformation. The *transition probabilities* associated with F are defined by

$$p(v,B) = \mu\big(\{x \in M : F_x(v) \in B\}\big) = \mu\big(F^{-1}(M \times B)_v\big),$$

for each $v \in N$ and each measurable set $B \subset N$. The set on the right-hand side is measurable (Exercise 5.1) and so $p(v,B)$ is well defined. It is clear that $p(v,B)$ is countably additive on the second variable, and so each $p(v,\cdot)$ is a probability measure on N. Moreover (Exercise 5.1), the function $N \to \mathbb{R}$, $v \mapsto p(v,B)$ is measurable, for any measurable $B \subset N$.

The *transition operator* $\mathscr{P}$ associated with the random transformation F is the linear map $\mathscr{P}$ acting on the space of bounded measurable functions $\varphi : N \to \mathbb{R}$ by

$$\mathscr{P}\varphi : N \to \mathbb{R}, \quad \mathscr{P}\varphi(v) = \int \varphi(F_x(v))\, d\mu(x). \tag{5.1}$$

Note that $x \mapsto \varphi(F_x(v))$ is measurable (Exercise 5.2), and so $\mathscr{P}\varphi(v)$ is well defined. It is clear that the function $\mathscr{P}\varphi$ defined by (5.1) is bounded. Moreover, $\mathscr{P}\varphi$ is measurable. To see this, begin by supposing that φ is the characteristic function of some set $B \subset N$. Then

$$\mathscr{P}\varphi(v) = \mu(\{x : F_x(v) \in B\}) = p(v, B) \tag{5.2}$$

is a measurable function of v, as observed in the previous paragraph. By the linearity of the transition operator, it follows that $\mathscr{P}\varphi$ is measurable whenever φ is a simple function; that is, a linear combination of characteristic functions. In general, there exists a sequence $(\varphi_n)_n$ of simple functions converging uniformly to φ. Thus, $\mathscr{P}\varphi$ is the uniform limit of measurable functions $\mathscr{P}\varphi_n$, $n \geq 1$, and so it is measurable.

The *adjoint transition operator* $\mathscr{P}^*$ associated with the random transformation F acts on the space of probability measures η on N by

$$\mathscr{P}^*\eta(B) = \int (F_x)_*\eta(B)\, d\mu(x) = \int \eta\left(F_x^{-1}(B)\right) d\mu(x), \tag{5.3}$$

for any measurable set $B \subset N$. Observe that

$$\eta\left(F_x^{-1}(B)\right) = \eta\left(F^{-1}(M \times B)^x\right) \tag{5.4}$$

is a measurable function of x, by Exercise 5.1. Thus, the integral in (5.3) is well defined. Moreover, it is clear from the definition (5.3) that $\mathscr{P}^*\eta$ is countably additive and satisfies $\mathscr{P}^*\eta(N) = 1$. Thus, $\mathscr{P}^*\eta$ is a probability measure on N.

We will take the space $\mathscr{T}(N)$ of measurable transformations $N \to N$ to be endowed with some σ-algebra relative to which the map $\mathscr{F} : M \to \mathscr{T}(N)$, $\mathscr{F}(x) = F_x$ is measurable. For instance, this could be the push-forward of the σ-algebra $\mathscr{A}$ under the map $\mathscr{F}$. Then we denote by ν the push-forward $\mathscr{F}_*\mu$ of the probability measure μ under $\mathscr{F}$. The definitions (5.1) and (5.3) translate to

$$\mathscr{P}\varphi(v) = \int \varphi(g(v))\, d\nu(g) \quad \text{and} \quad \mathscr{P}^*\eta(B) = \int \eta(g^{-1}(B))\, d\nu(g). \tag{5.5}$$

So, the transition operators are completely characterized by the probability measure ν. We have the following duality relation:

Lemma 5.3 *Let $\varphi : N \to \mathbb{R}$ be a bounded measurable function. Then*

$$\int \varphi\, d(\mathscr{P}^*\eta) = \int (\mathscr{P}\varphi)\, d\eta.$$

Proof Suppose first that φ is the characteristic function of some measurable set $B \subset N$. Then, using (5.4),

$$\int \varphi\, d(\mathscr{P}^*\eta) = \mathscr{P}^*\eta(B) = \int \eta\left(F^{-1}(M \times B)^x\right) d\mu(x) = (\mu \times \eta)(F^{-1}(M \times B))$$

and, using (5.2),

$$\int (\mathscr{P}\varphi)\,d\eta = \int \mu(\{x : F_x(v) \in B\})\,d\eta(v) = (\mu \times \eta)(F^{-1}(M \times B)).$$

This proves the equality in this case. By linearity, we get that the equality holds for every simple function φ. The general case follows, because every bounded measurable function is a uniform limit of simple functions. □

5.2 Stationary measures

A probability measure η on N is called *stationary* for the random transformation F if $\mathscr{P}^*\eta = \eta$; that is, if

$$\eta(B) = \int \eta(F_x^{-1}(B))\,d\mu(x) = \int \eta(g^{-1}(B))\,d\nu(g) \tag{5.6}$$

for every measurable set $B \subset N$. The probability measure $\nu = \mathscr{F}_*\mu$ was introduced in Section 5.1. The expression on the right-hand side of (5.6) shows that ν determines the set of stationary measures entirely.

Recall that a measure η is said to be *invariant* under a transformation g if $\eta(g^{-1}(B)) = \eta(B)$ for every measurable set. The definition (5.6) means that η is called stationary if such an equality holds *on average* over all F_x, the average being with respect to μ. Thus, clearly, any probability measure that is invariant under F_x for μ-almost every x is also stationary. The following example (see also Exercise 6.8) illustrates the fact that the converse is far from being true.

Example 5.4 Let $f : M \to M$ be the shift map on $M = \{1,2\}^{\mathbb{Z}}$ and μ be a Bernoulli measure supported on the whole $M = \{1,2\}^{\mathbb{Z}}$. Let $A : M \to \mathrm{SL}(2)$ be a locally constant function,

$$A \mid [0;1] \equiv A_1 \quad \text{and} \quad A \mid [0;2] \equiv A_2,$$

where A_1 and A_2 are hyperbolic matrices with no common eigenspace. Let $F : M \times \mathbb{PR}^2 \to M \times \mathbb{PR}^2$ be the projective cocycle defined by A over f. For each $i = 1,2$, the probability measures on $\mathbb{PR}^2$ invariant under the action of A_i are the convex combinations of the Dirac masses on the corresponding eigenspaces. Thus, the hypotheses imply that A_1 and A_2 have no common invariant probability measure. On the other hand, Proposition 5.6 below ensures that F does have some stationary measure.

The following characterization of stationary measures is specific to *one-sided random transformations*; that is, such that $M = X^{\mathbb{N}}$ and $f : M \to M$ is the one-sided shift. The two-sided situation will be analyzed later.

Proposition 5.5 *Let $F : M \times N \to M \times N$ be a one-sided random transformation. A probability measure η on N is stationary for F if and only if the probability measure $\mu \times \eta$ on $M \times N$ is F-invariant.*

Proof First, we prove the 'if' claim. Suppose that $\mu \times \eta$ is invariant under F. Given any bounded measurable function $\varphi : N \to \mathbb{R}$, define $\psi : M \times N \to \mathbb{R}$ by $\psi(x,v) = \varphi(v)$. Then, using Lemma 5.3,

$$\begin{aligned}\int \varphi(v)(d\mathscr{P}^*\eta)(v) = \int \mathscr{P}\varphi(v)\,d\eta(v) &= \int\int \psi(x,F_x(v))\,d\mu(x)\,d\eta(v)\\ &= \int\int \psi(f(x),F_x(v))\,d\mu(x)\,d\eta(v)\\ &= \int\int \psi(x,v)\,d\mu(x)\,d\eta(v) = \int \varphi(v)\,d\eta(v).\end{aligned}$$

Since φ is arbitrary, this shows that $\mathscr{P}^*\eta = \eta$, as we wanted to prove.

Now we prove the 'only if' claim. Let η be a stationary measure on N. Given any bounded measurable function $\psi : M \times N \to \mathbb{R}$, consider the function $\varphi : N \to \mathbb{R}$ defined by $\varphi(v) = \int \psi(x,v)d\mu(x)$. Then

$$\begin{aligned}\int\int \psi(x,v)\,d\mu(x)\,d\eta(v) = \int \varphi(v)\,d\eta(v) &= \int \mathscr{P}\varphi(v)\,d\eta(v)\\ &= \int\int \varphi(F_x(v))\,d\mu(x)\,d\eta(v)\\ &= \int\int\int \psi(y,F_x(v))\,d\mu(y)\,d\mu(x)\,d\eta(v).\end{aligned}$$

Write $x = (x_0, x_1, \dots)$ and $y = (y_0, y_1, \dots)$. Since F_x depends only on x_0, the triple integral may be written as

$$\int\int\int \psi(y,F_{x_0}(v))\,dp^{\mathbb{N}}(y)\,dp(x_0)\,d\eta(v).$$

Let us write $z = (x_0, y_0, y_1, \dots)$. Then $f(z) = y$, and so the previous integral becomes

$$\int\int \psi(f(z),F_{z_0}(v))\,dp^{\mathbb{N}}(z)\,d\eta(v) = \int\int \psi(f(z),F_z(v))\,d\mu(z)\,d\eta(v).$$

This proves that

$$\int\int \psi\,d\mu\,d\eta = \int\int (\psi \circ F)\,d\mu\,d\eta.$$

Since ψ is arbitrary, this shows that $\mu \times \eta$ is invariant under F. □

There is more than one useful topology in the space of probability measures on a measurable space N. The *uniform topology* is defined by the *total variation norm*

$$\|\xi - \eta\| = \sup_{|\psi|\le 1} \left| \int \psi \, d\xi - \int \psi \, d\eta \right|,$$

where the supremum is over all measurable functions $\psi : M \to \mathbb{R}$ with $|\psi| \le 1$. The *pointwise topology* is the smallest topology such that

$$\xi \mapsto \xi(E) \quad \text{is continuous for every measurable } E \subset N.$$

It follows that $\xi \mapsto \int \psi \, d\xi$ is continuous for every bounded measurable function $\psi : M \to \mathbb{R}$. When N is a metric space, we may also consider the *weak* topology*, which is the smallest topology such that

$$\xi \mapsto \int \varphi \, d\xi \text{ is continuous for each bounded uniformly continuous function } \varphi.$$

It is clear from the definitions that uniform convergence implies pointwise convergence which, in turn, implies weak* convergence. Both converses are false, in general. The weak* topology is especially useful, not the least because it is compact if N is compact (see [114, Theorem 2.1.5]).

Proposition 5.6 *Assume that N is a compact metric space and $F_x : N \to N$ is continuous for every $x \in M$. Then there is some stationary measure for F.*

Proof The first step is

Lemma 5.7 *If $\varphi : N \to \mathbb{R}$ is continuous then $\mathscr{P}\varphi : N \to \mathbb{R}$ is also continuous.*

Proof Given $n \in \mathbb{N}$, $v \in N$, and $\delta > 0$, define

$$\mathscr{B}(v,n,\delta) = \left\{ x \in M : F_x\left(B\left(v, \frac{1}{n}\right)\right) \subset B(F_x(v), \delta) \right\}.$$

Fix $v \in N$ and $\varepsilon > 0$. Then fix $\delta > 0$ such that

$$d(z,y) < \delta \Rightarrow |\varphi(z) - \varphi(y)| < \frac{\varepsilon}{2}.$$

The fact that every F_x is continuous ensures that $\mu(\mathscr{B}(v,n,\delta)^c) \to 0$ as $n \to \infty$. Fix n large enough so that

$$\mu(\mathscr{B}(v,n,\delta)^c) < \frac{\varepsilon}{4 \sup |\varphi|}.$$

Now we are ready to prove continuity. For $w \in N$ such that $d(v,w) < \delta$,

$$\begin{aligned}|\mathscr{P}\varphi(v) - \mathscr{P}\varphi(w)| &= \left|\int [\varphi(F_x(v)) - \varphi(F_x(w))]\, d\mu(x)\right| \\ &\le \int_{\mathscr{B}(v,n,\delta)} |\cdots|\, d\mu(x) + \int_{\mathscr{B}(v,n,\delta)^c} |\cdots|\, d\mu(x) \\ &\le \frac{\varepsilon}{2}\mu(\mathscr{B}(v,n,\delta)) + 2\sup|\varphi|\mu(\mathscr{B}(v,n,\delta)^c) \le \varepsilon\end{aligned}$$

This proves the lemma. □

Lemma 5.8 *The operator $\mathscr{P}^*$ is continuous relative to the weak* topology in the space of probability measures on N.*

Proof Let $(\eta_k)_k$ be a sequence of probability measures on N converging to η in the weak* topology. Let $\varphi : N \to \mathbb{R}$ be continuous. By Lemma 5.7, the function $\mathscr{P}\varphi$ is also continuous. Then, using Lemma 5.3 and the definition of the weak* topology,

$$\int \varphi\, d(\mathscr{P}^*\eta_k) = \int (\mathscr{P}\varphi)\, d\eta_k \to \int (\mathscr{P}\varphi)\, d\eta = \int \varphi d(\mathscr{P}^*\eta).$$

This proves that $\mathscr{P}^*\eta_k$ converges to $\mathscr{P}^*\eta$ in the weak* topology, and so $\mathscr{P}^*$ is continuous. □

To finish the proof of Proposition 5.6, let ξ be an arbitrary probability measure on N and then define

$$\eta_n = \frac{1}{n}\sum_{j=0}^{n-1}(\mathscr{P}^{*j}\xi).$$

Since the space of probability measures on N is weak* compact, there exists $(n_k)_k \to \infty$ such that η_{n_k} converges in the weak* topology to some probability measure η. On the one hand, by Lemma 5.8, $\mathscr{P}^*\eta_{n_k}$ converges to $\mathscr{P}^*\eta$. On the other hand,

$$\mathscr{P}^*\eta_k = \frac{1}{n_k}\sum_{j=1}^{n}\mathscr{P}^{*j}\eta_j, = \eta_k + \frac{1}{n_k}\left(\mathscr{P}^{*n_k}\xi - \xi\right)$$

and so $\mathscr{P}^*\eta_k$ also converges to η. These two facts show that $\mathscr{P}^*\eta = \eta$ as we wanted to prove. □

Since the stationary measures are the fixed points of $\mathscr{P}^*$, it also follows from Lemma 5.8 that, under the assumptions of Proposition 5.6, the set of stationary measures for F is closed (hence, compact) for the weak* topology. In fact, one has a stronger statement, where the cocycle and the Bernoulli measure are allowed to vary: the set of triples (p, F, η) such that η is a stationary measure is closed.

To state this in precise terms, let $(p_k)_k$ be a sequence of probability measures on X, let $\mu_k = p_k^{\mathbb{N}}$ (or $p_k^{\mathbb{Z}}$) for each $k \geq 1$, let $F_k : M \times N \to M \times N$, $k \geq 1$ be random transformations over the Bernoulli shifts $f : (M, \mu_k) \to (M, \mu_k)$, and let η_k be a stationary measure for each F_k, $k \geq 1$.

Proposition 5.9 *Take N to be a compact metric space and $F_x : N \to N$ to be continuous for every $x \in M$.*

(a) *Suppose that $(p_k)_k$ converges to p in the pointwise topology, $(F_k)_k$ converges to F uniformly on $M \times N$, and $(\eta_k)_k$ converges to η in the weak* topology. Then η is a stationary measure for F.*

(b) *If X is a metric space and $x \mapsto F_x(v)$ is continuous for every $v \in N$, then the previous conclusion remains true when $(p_k)_k$ converges to p in the weak* topology and the other assumptions are unchanged.*

Proof The argument is essentially the same for both claims, pointwise and weak*. For each $k \geq 1$, let $\mathscr{P}_k$ and $\mathscr{P}_k^*$ be the transition operators associated with the random transformation F_k.

Lemma 5.10 *$\mathscr{P}_k\varphi$ converges uniformly to $\mathscr{P}\varphi$, for any continuous $\varphi : N \to \mathbb{R}$.*

Proof Let $\varepsilon > 0$. By definition, for any $v \in N$,

$$\mathscr{P}_k\varphi(v) - \mathscr{P}\varphi(v) = \int \varphi(F_{k,x}(v))\, dp_k(x) - \int \varphi(F_x(v))\, dp_k(x) + \int \varphi(F_x(v))\, dp_k(x) - \int \varphi(F_x(v))\, dp(x). \tag{5.7}$$

Since φ is (uniformly) continuous and F_k is uniformly close to F,

$$|\varphi(F_{k,x}(v)) - \varphi(F_x(v))| < \varepsilon \quad \text{for every } (x, v) \in M \times N$$

if k is large enough. Then the first term of (5.7) on the right-hand side of (5.7) is smaller than ε. The last term of (5.7) is also smaller than ε if k is large, because $(p_k)_k$ converges to p and the function $x \mapsto \varphi(F_x(v))$ is bounded and measurable (respectively, continuous). This proves the lemma. □

Proceeding with the proof of Proposition 5.9, consider any sequence $(\eta_k)_k$ converging to some η. For any continuous function $\varphi : N \to \mathbb{R}$,

$$\int \varphi\, d(\mathscr{P}_k^*\eta_k) - \int \varphi\, d(\mathscr{P}^*\eta) = \\ = \int (\mathscr{P}_k\varphi)\, d\eta_k - \int (\mathscr{P}\varphi)\, d\eta_k + \int \varphi\, d(\mathscr{P}^*\eta_k) - \int \varphi\, d(\mathscr{P}^*\eta).$$

By Lemma 5.10, the first difference on the right-hand side goes to zero when $k \to \infty$. The second difference goes to zero as well because, by Lemma 5.8, $\mathscr{P}^*\eta_k$ converges to $\mathscr{P}^*\eta$ in the weak* topology. Since φ is arbitrary, this proves that $\mathscr{P}_k^*\eta_k$ converges to $\mathscr{P}^*\eta$ in the weak* topology, as $k \to \infty$. When each η_k is stationary for F_k this means that η_k converges to $\mathscr{P}^*\eta$. Then $\mathscr{P}^*\eta = \eta$, as we wanted to prove. □

5.3 Ergodic stationary measures

A bounded measurable function $\varphi : N \to \mathbb{R}$ is *stationary* if it satisfies $\mathscr{P}\varphi = \varphi$. A set $B \subset N$ is *stationary* if its characteristic function $\mathscr{X}_B$ is stationary. Let η be a stationary measure. A bounded measurable function $\varphi : N \to \mathbb{R}$ is η-*stationary* if $\mathscr{P}\varphi(v) = \varphi(v)$ for η-almost every $v \in N$. A set $B \subset N$ is η-*stationary* if $\mathscr{X}_B$ is η-stationary.

Proposition 5.11 *Let η be a stationary measure. The following conditions are equivalent:*

(a) *every η-stationary function is constant on some set with full η-measure;*
(b) *if $B \subset N$ is an η-stationary set then $\eta(B)$ is either* 0 *or* 1.

Proof Property (b) is a special case of property (a). To prove that (b) also implies (a), let $\varphi : N \to \mathbb{R}$ be η-stationary. We claim that

$$B(c) = \{v \in N : \varphi(v) > c\}$$

is an η-stationary set for every $c \in \mathbb{R}$. Consequently, $\eta(B(c))$ is either 0 or 1, for every $c \in \mathbb{R}$. Let $\bar{c}$ be the supremum of the set of values of c such that $\eta(B(c)) = 1$. Then $\varphi(v) = \bar{c}$ for η-almost every $v \in B(\bar{c})$. We are left to prove the claim.

Lemma 5.12 *If η is a stationary measure and $\psi_1, \psi_2 : N \to \mathbb{R}$ are η-stationary then so are the functions* $\max\{\psi_1, \psi_2\}$ *and* $\min\{\psi_1, \psi_2\}$.

Proof Begin by noting that if ψ is η-stationary then $|\psi| = |\mathscr{P}\psi| \le \mathscr{P}|\psi|$ at η-almost every point and $\int (\mathscr{P}|\psi| - |\psi|)\, d\eta = 0$, because η is stationary. These two facts together imply that $|\psi|$ is η-stationary. It is clear that the sum and the difference of η-stationary functions are also η-stationary. Then, for any ψ_1 and ψ_2 as in the statement

$$\max\{\psi_1, \psi_2\} = \frac{\psi_1 + \psi_2}{2} + \frac{|\psi_1 - \psi_2|}{2}$$

and

$$\min\{\psi_1, \psi_2\} = \frac{\psi_1 + \psi_2}{2} - \frac{|\psi_1 - \psi_2|}{2}$$

are η-stationary. This proves lemma. □

It follows from Lemma 5.12 that $\varphi_n = \min\{1, n\max\{\varphi - c, 0\}\}$ is η-stationary for every $n \geq 1$. Now, φ_n converges monotonically to the characteristic function of $B(c)$. So (Exercise 5.7), the characteristic function is η-stationary, as claimed. This finishes the proof of the proposition. □

A stationary measure η is *ergodic* if it satisfies conditions (a) and (b) in Proposition 5.11. It follows from Exercise 5.6 that the definition does not change if one considers, instead, sets or functions that are actually stationary (not just η-stationary).

In the one-sided case, we have the following complement to Proposition 5.5:

Proposition 5.13 *Let $F : M \times N \to M \times N$ be a one-sided random transformation and η be a stationary measure. Then η is ergodic if and only if the F-invariant probability measure $m = \mu \times \eta$ is ergodic.*

Proof Suppose that $m = \mu \times \eta$ is ergodic and let $B \subset N$ be a stationary set. The latter means that

$$\mathscr{X}_B(v) = \mathscr{P}\mathscr{X}_B(v) = \int \mathscr{X}_B(F_x(v))\, d\mu(x)$$

for every $v \in N$. This may be read as follows: $v \in B$ implies $F_x(v) \in B$ for μ-almost every x and $v \notin B$ implies $F_x(v) \notin B$ for μ-almost every x. Then the set $M \times B$ is (F, m)-invariant, in the sense that the symmetric difference $(M \times B) \Delta F^{-1}(M \times B)$ has zero m-measure. By ergodicity, it follows that $\eta(B) = m(M \times B)$ is either 0 or 1. This proves the "if" part of the statement.

To prove the "only if" part, let $\psi : M \times N \to \mathbb{R}$ be any bounded measurable F-invariant function; that is, such that $\psi \circ F = \psi$. Let $\varphi : N \to \mathbb{R}$ be defined by $\varphi(v) = \int \psi(y, v)\, d\mu(y)$. Then, for any $v \in N$,

$$\mathscr{P}\varphi(v) = \int \varphi(F_x(v))\, d\mu(x) = \int \int \psi(y, F_x(v))\, d\mu(y)\, d\mu(x).$$

By assumption, F_x depends only on the zeroth coordinate x_0. On the other hand, $f(x) = (x_1, \ldots, x_n, \ldots)$ is independent of x_0. Thus, recalling that $\mu = p^{\mathbb{N}}$, the last integral coincides with

$$\int \psi(f(x), F_x(v))\, d\mu(x) = \int \psi(x, v)\, d\mu(x) = \varphi(v)$$

for every $v \in N$. This means that φ is stationary. Hence, by hypothesis, there exists $C \in \mathbb{R}$ such that $\varphi(v) = C$ for η-almost every v. Moreover, for each $k \geq 1$ and $x_0, x_1, \ldots, x_{k-1} \in X$,

$$\int \psi(x,v)\, d\mu(x_k, x_{k+1}, \ldots) = \int \psi(f^k(x), F_x^k(v))\, d\mu(x_k, x_{k+1}, \ldots).$$

Changing variables $y = f^k(x) = (x_k, x_{k+1}, \ldots)$, and observing that F_x^k depends only on $x_0, \ldots, x_{k-1}$, we conclude that

$$\int \psi(x,v)\, d\mu(x_k, x_{k+1}, \ldots) = \int \psi(y, F_x^k(v))\, d\mu(y) = \varphi(F_x^k(v)).$$

This proves that

$$\int \psi(x,v)\, d\mu(x_k, x_{k+1}, \ldots) = C \quad \text{for every } k, x_0, \ldots, x_{k-1}, v.$$

Then (Exercise 5.8), we have $\psi(x,v) = C$ for m-almost all $(x,v) \in M \times N$. This proves that m is ergodic, and so the proof of the proposition is complete. □

The next theorem is an analogue for random transformations of the ergodic decomposition theorem in deterministic dynamics (see [114, Theorem 5.1.3]): every stationary measure can be written as a convex combination of ergodic stationary measures.

Theorem 5.14 (Ergodic decomposition) *Take N to be a separable complete metric space. Then the space of ergodic stationary measures is a measurable subset of the space of all stationary measures. Moreover, every stationary measure η may be represented as $\eta = \int \xi \, d\hat{\eta}(\xi)$, meaning that*

$$\eta(B) = \int \xi(B)\, d\hat{\eta}(\xi) \quad \textit{for every Borel set } B \subset N,$$

where $\hat{\eta}$ is a probability measure in the space of stationary measures giving full weight to the subset of ergodic stationary measures.

See Kifer [71, Appendix A.1] for a proof. The statement extends, immediately, to the case when N is just a Borel subset of a separable complete metric space, because any measure on N may be identified with a measure on the larger metric space.

5.4 Invertible random transformations

We have seen in Proposition 5.5 that, in the one-sided case, a probability measure η is stationary if and only if the probability measure $\mu \times \eta$ is invariant. Here we discuss the relations between these two concepts when the random transformation F is invertible and, in particular, the shift $f : M \to M$ is two-sided.

Throughout, we assume that X is a separable complete metric space. Then (Exercise 5.12), $M = X^{\mathbb{Z}}$ (and $M = X^{\mathbb{N}}$) itself is also a separable complete metric space.

Let $f : (M, \mu) \to (M, \mu)$ be a two-sided Bernoulli shift: $M = X^{\mathbb{Z}}$ and the invariant probability measure $\mu = p^{\mathbb{Z}}$ for some probability measure p on X. Denote

$$\mathbb{Z}^+ = \{n \in Z : n \geq 0\} \quad \text{and} \quad \mathbb{Z}^- = \{n \in \mathbb{Z} : n < 0\}.$$

Then write $M^{\pm} = X^{\mathbb{Z}^{\pm}}$ and $\mu^{\pm} = p^{\mathbb{Z}^{\pm}}$. Let $\pi^{\pm} : M \to M^{\pm}$ be the canonical projections, and define $f^{\pm} : M^{\pm} \to M^{\pm}$ by

$$f^+ \circ \pi^+ = \pi^+ \circ f \quad \text{and} \quad f^- \circ \pi^- = \pi^- \circ f^{-1}.$$

Then $f^+ : (M^+, \mu^+) \to (M^+, \mu^+)$ and $f^- : (M^-, \mu^-) \to (M^-, \mu^-)$ are one-sided Bernoulli shifts and $\mu = \mu^- \times \mu^+$.

Now let $F : M \times N \to M \times N$ be an invertible random transformation over $f : (M, \mu) \to (M, \mu)$. There are two non-invertible random transformations naturally associated with F, namely:

$$\begin{aligned} F^+ &: M^+ \times N \to M^+ \times N, \quad F^+(x^+, v) = (f^+(x^+), F^+_{x^+}(v)) \quad \text{over } f^+ \\ F^- &: M^- \times N \to M^- \times N, \quad F^-(x^-, v) = (f^-(x^-), F^-_{x^-}(v)) \quad \text{over } f^-, \end{aligned}$$

where $F^+_{x^+} = F_x$ for any $x \in M$ with $\pi^+(x) = x^+$ and $F^-_{x^-} = \left(F_{f^{-1}(x)}\right)^{-1}$ for any $x \in M$ with $\pi^-(x) = x^-$. Since $F_x : N \to N$ depends only on the zeroth coordinate x_0 of the point $x \in M$, these transformations are indeed well defined and locally constant: $F^+_{x^+}$ depends only on the coordinate x^+_0 and $F^-_{x^-}$ depends only on the coordinate x^-_{-1}.

A probability measure η on N is *forward stationary* for F if it is stationary for F^+; that is, if

$$\eta(B) = \int \eta\left((F^+_{x^+})^{-1}(B)\right) d\mu^+(x^+) = \int \eta\left(F_x^{-1}(B)\right) d\mu(x)$$

for every measurable set B. Analogously, η is *backward stationary* for F if it is stationary for F^-; that is, if

$$\begin{aligned} \eta(B) &= \int \eta\left((F^-_{x^-})^{-1}(B)\right) d\mu^-(x^-) = \int \eta\left(F_{f^{-1}(x)}(B)\right) d\mu(x) \\ &= \int \eta\left(F_x(B)\right) d\mu(x) \end{aligned}$$

for every measurable set B (the last equality uses the fact that the measure μ is invariant under f).

In what follows we are going to analyze these notions from several different angles. The conclusions can be summarized as follows (some of the notions involved have yet to be defined):

$$\begin{aligned}
\eta \text{ is forward stationary} &\Leftrightarrow \text{the product } m^+ = \mu^+ \times \eta \text{ is } F^+\text{-invariant}\\
&\Leftrightarrow \text{the lift } m \text{ of } \eta \text{ (and } m^+\text{) is a } u\text{-state}\\
&\Leftrightarrow \text{the disintegration of } m \text{ factors through } M^-\\
\eta \text{ is backward stationary} &\Leftrightarrow \text{the product } m^- = \mu^- \times \eta \text{ is } F^-\text{-invariant}\\
&\Leftrightarrow \text{the lift } m \text{ of } \eta \text{ (and } m^-\text{) is an } s\text{-state}\\
&\Leftrightarrow \text{the disintegration of } m \text{ factors through } M^+.
\end{aligned}$$

In general, the sets of forward stationary measures and backward stationary measures are distinct and even disjoint. That is the case, for instance, for locally constant projective cocycles that are pinching and twisting (Exercise 5.24).

5.4.1 Lift of an invariant measure

The invariant probabilities of F over μ are in one-to-one correspondence with the invariant probabilities of F^+ over μ^+, and the same is true for F^- and μ^-. To state this fact precisely, consider the canonical projections

$$\begin{aligned}
P: M\times N \to M \quad &\text{and} \quad Q: M\times N \to N\\
P^\pm: M^\pm \times N \to M^\pm \quad &\text{and} \quad Q^\pm: M^\pm \times N \to N,
\end{aligned} \tag{5.8}$$

and let $\Pi^\pm: M\times N \to M^\pm \times N$ be given by $\Pi^\pm(x,v) = (\pi^\pm(x), v)$. Recall that $\pi^\pm: M \to M^\pm$ are the canonical projections between the shift spaces. The next two lemmas remain true if one replaces all $+$-signs by $-$-signs.

Lemma 5.15

(1) *If m is an F-invariant probability measure with $P_* m = \mu$ then $m^+ = \Pi^+_* m$ is an F^+-invariant probability measure with $P^+_* m^+ = \mu^+$.*
(2) *Given any F^+-invariant probability measure m^+ with $P^+_* m^+ = \mu^+$ there exists a unique F-invariant probability measure m such that $\Pi^+_* m = m^+$ and $P_* m = \mu$.*

Proof The first claim is an immediate consequence of the obvious relations $F^+ \circ \Pi^+ = \Pi^+ \circ F$ and $P^+ \circ \Pi^+ = \pi^+ \circ P$. To prove the second claim, let $\mathscr{C}$ be the σ-algebra of $M \times N$ and, for each $n \geq 0$, denote by $\mathscr{C}_n$ the sub-σ-algebra of sets of the form $F^n(M^- \times G)$ for some $G \subset M^+ \times N$. Note that

$$\mathscr{C}_0 \subset \mathscr{C}_1 \subset \cdots \subset \mathscr{C}_n \subset \cdots \quad \text{and} \quad \bigcup_{n=0}^{\infty} \mathscr{C}_n \text{ generates } \mathscr{C}. \tag{5.9}$$

Suppose that m is an F-invariant probability measure such that $\Pi_*^+ m = m^+$. Then

$$m(F^n(M^- \times G)) = m(M^- \times G) = m^+(G) \tag{5.10}$$

for any measurable set $G \subset M^+ \times N$. This proves that m is uniquely determined on $\mathscr{C}_n$ for every n. Then, by (5.9), m is uniquely determined on $\mathscr{C}$. To prove existence, take (5.10) to define m on each σ-algebra $\mathscr{C}_n$. These definitions are consistent: if

$$F^n(M^- \times G_n) = F^{n-1}(M^- \times G_{n-1})$$

then

$$M^- \times G_n = F^{-1}(M^- \times G_{n-1}) = M^- \times (F^+)^{-1}(G_{n-1})$$

and, using that m^+ is F_+-invariant, we find that $m^+(G_{n-1}) = m^+(G_n)$. So, (5.10) does define a probability measure on $\mathscr{C}$. By construction, this probability measure is F-invariant. □

The probability measure m is called the *lift* of m^+. Note that the operations in parts (1) and (2) of Lemma 5.15 are inverse to each other: by definition, the Π_*^+-projection of the lift of m^+ coincides with m^+ and, by uniqueness, the lift of the Π^+-projection of m coincides with m.

Lemma 5.16 *An F^+-invariant probability measure m^+ is ergodic for F^+ if and only if its lift m is ergodic for F.*

Proof Let $G \subset M^+ \times N$ be a measurable set invariant under F^+. Then $M^- \times G$ is invariant under F. If m is ergodic, it follows that $m^+(G) = m(M^- \times G)$ is either 0 or 1. Thus, m^+ is ergodic.

To prove the converse, let $\psi : M \times N \to \mathbb{R}$ be a bounded measurable function. We claim that the Birkhoff time average, relative to the transformation F, is constant m-almost everywhere. It follows that, m is ergodic for F. To prove the claim, suppose first that there exists $k \geq 1$ such that $\psi(x, v)$ depends only on $(x_{-k}, \ldots, x_0, \ldots)$ and $v \in N$. Then, for every $n \geq k$

$$\psi(F^n(x, v)) = \psi(f^n(x), F_x^n(v))$$

depends only on $x^+ = (x_0, \ldots, x_n, \ldots)$ and v. Consequently, the time average $\tilde{\psi}$ is constant on each set $M^- \times \{(x^+, v)\}$. This means that $\tilde{\psi} : M \times N \to \mathbb{R}$ may be identified with a function $\hat{\psi} : M^+ \times N \to \mathbb{R}$, defined on a subset with full

m^+-measure. Moreover, the fact that $\tilde{\psi} \circ F = \tilde{\psi}$ at m-almost every point entails that $\hat{\psi} \circ F^+ = \hat{\psi}$ at m^+-almost every point. By ergodicity of m^+, it follows that $\hat{\psi}$ is constant on a subset with full m^+-measure. Equivalently, $\tilde{\psi}$ is constant on a subset with full m-measure. This proves the claim in this case. In general, for each $k \geq 1$, let $\psi_k : M \times N \to \mathbb{R}$ be defined by

$$\psi_k(x,v) = \int \psi(x,v)\, d\mu(\ldots,x_{-m},\ldots,x_{-k+1}).$$

By the previous special case of our claim, the time average $\tilde{\psi}_k$ is constant m-almost everywhere. Now, $(\psi_k)_k$ converges to ψ at m-almost every point (Exercise 5.16). By the bounded convergence theorem, it follows that $(\psi_k)_k$ converges to ψ in the space $L^1(m)$. Then the time averages $(\tilde{\psi}_k)_k$ also converge to $\tilde{\psi}$ in $L^1(m)$. It follows that $\tilde{\psi}$ is constant m-almost everywhere. □

5.4.2 s-states and u-states

In view of the previous observations, to every forward stationary measure η one can associate an F-invariant probability measure m, namely the lift of the F^+-invariant probability measure $m^+ = \mu^+ \times \eta$; similarly for every backward stationary measure. In either case, we call m the *lift* of the stationary measure η. Our purpose now is to characterize the classes of F-invariant measures one finds in this way.

Let m be an F-invariant probability measure on $M \times N$ with $P_* m = \mu$. We call m an *s-state* if, for measurable sets $A^- \subset M^-$, $A^+ \subset M^+$, and $B \subset N$,

$$\frac{m(A^- \times A^+ \times B)}{\mu(A^- \times A^+)} \quad \text{does not depend on } A^-. \tag{5.11}$$

Analogously, m is called a *u-state* if, for measurable sets $A^- \subset M^-$, $A^+ \subset M^+$, and $B \subset N$,

$$\frac{m(A^- \times A^+ \times B)}{\mu(A^- \times A^+)} \quad \text{does not depend on } A^+. \tag{5.12}$$

We call m an *su-state* if it is both an s-state and a u-state. These are measures of a very special kind:

Proposition 5.17 *If m is a u-state then $\eta = Q_* m$ is a forward stationary measure. Conversely, if η is a forward stationary measure on N then its lift m is a u-state. Moreover, m is ergodic if and only if η is ergodic. The same is true replacing, simultaneously, forward stationary by backward stationary and u-state by s-state.*

Proof Let m be a u-state and $m^+ = \Pi^+_* m$. For any measurable sets $A^+ \subset M^+$ and $B \subset N$,

$$\frac{m^+(A^+ \times B)}{\mu^+(A^+)} = \frac{m(M^- \times A^+ \times B)}{\mu(M^- \times A^+)}$$

depends only on B. Let this expression be denoted $\eta(B)$. It is clear from the definition that η is a probability measure on N. Moreover, $m^+ = \mu^+ \times \eta$ and, since m^+ is invariant under F^+, Proposition 5.5 asserts that η is forward stationary for F. Finally, $Q_* m = Q^+_* m^+ = \eta$. This proves the first claim.

To prove the converse, suppose η is a stationary measure and let m be its lift. We want to prove that

$$\frac{m(A^- \times A^+ \times B)}{\mu(A^- \times A^+)}$$

does not depend on A^+. It is no restriction to take A^- to be a cylinder,

$$A^- = \{x^- \in M^- : x_{-k} \in X_{-k}, \dots, x_{-1} \in X_{-1}\},$$

for some measurable sets X_{-k}, …, $X_{-1} \subset X$, because the cylinders generate the σ-algebra of M^-. Then, since m is invariant under F,

$$\begin{aligned} m(A^- \times A^+ \times B) &= m(F^{-k}(A^- \times A^+ \times B)) \\ &= m(M^- \times G) = (\mu^+ \times \eta)(G) \end{aligned}$$

where G is the set of $(x^+, v) \in M^+ \times N$ with $x_0 \in X_{-k}$, …, $x_{k-1} \in X_{-1}$, $(x_k, \dots, x_n, \dots) \in A^+$, and $v \in (F^k_x)^{-1}(B)$. Notee that F^k_x is a function of x^+, indeed it depends only on x_0, …, x_{k-1}. We conclude that $m(A^- \times A^+ \times B)$ is given by

$$\int_{X_{-k} \times \cdots \times X_{-1}} \mu^+(A^+) \eta((F^k_x)^{-1}(B)) \, dp^k(x_0, \dots, x_{k-1}).$$

Consequently,

$$\frac{m(A^- \times A^+ \times B)}{\mu(A^- \times A^+)} = \frac{\int_{X_{-k} \times \cdots \times X_{-1}} \eta((F^k_x)^{-1}(B)) \, dp^k(x_0, \dots, x_{k-1})}{\mu^-(A^-)}$$

does not depend on A^+, as we wanted to prove. By Lemma 5.16, m is ergodic if and only if m^+ is ergodic. By Proposition 5.13, m^+ is ergodic if and only if η is ergodic. □

For the remainder of this section we take N to be a compact metric space, and $F_x : N \to N$ to be a homeomorphism for every $x \in M$. Moreover, $x_0 \mapsto F_{x_0}$ is a measurable map from X to the space of continuous transformations on N, endowed with the topology of uniform convergence. Keep in mind that we take X and M to be separable complete metric spaces.

It follows from Propositions 5.6 and 5.17 that the set of u-states (and s-states) is non-empty. Exercise 5.19 outlines an alternative, more direct proof of this fact, and similar ideas will be developed in Section 6.1.

The following stronger statement will be useful later. Let $(p_k)_k$ be a sequence of probability measures on X and $(\mu_k)_k$ be the corresponding Bernoulli measures on M. Let $F_k : M \times N \to M \times N$, $k \geq 1$ be random transformations over the Bernoulli shifts $f : (M, \mu_k) \to (M, \mu_k)$ and, for each $k \geq 1$, let m_k be an s-state (respectively, a u-state) for F_k.

Proposition 5.18 *Suppose that $(p_k)_k$ converges to p in the uniform topology, $(F_k)_k$ converges to F uniformly on $M \times N$, and $(m_k)_k$ converges to m in the weak* topology. Then m is an s-state (respectively, a u-state) for F.*

Moreover, the assumption on $(p_k)_k$ is not necessary when F is continuous.

Proof Since each m_k projects to μ_k and $(m_k)_k \to m$ in the weak* topology, the projections $(\mu_k)_k$ must converge to the projection of m in the weak* topology. Now, the hypothesis $(p_k)_k \to p$ implies that $(\mu_k)_k \to \mu$ in the weak* topology (Exercise 5.14). Thus, m projects down to μ.

Lemma 5.19 *Let $(m_k)_k$ be a sequence of probability measures on $M \times N$ projecting to Bernoulli measures $(\mu_k)_k$ on M and satisfying* (5.11). *If $(m_k)_k$ converges to some probability measure m and $(\mu_k)_k$ converges to some Bernoulli measure μ, in the weak* topology, then m satisfies* (5.11). *Analogously for condition* (5.12).

Proof Write $m_k = \mu_k^- \times m_k^+$ where m_k^+ is the projection of m_k to $M^+ \times N$ (Exercise 5.17). The hypothesis gives that $(m_k)_k \to m$ and $(\mu_k^-)_k \to \mu^-$ in the weak* topology. Restricting to some subsequence, we may assume that $(m_k^+)_k$ converges to some probability measure m^+ on $M^+ \times N$. Then (Exercise 5.14) $m = \mu^- \times m^+$ and this means that m satisfies (5.11). Exchanging $\pm$-signs, one gets the same claim for (5.12). □

To finish the proof of Proposition 5.18, we still have to prove that m is F-invariant. We claim that

$$\int (\varphi \circ F_k)\, dm_k - \int (\varphi \circ F)\, dm \to 0 \quad \text{as } k \to \infty, \tag{5.13}$$

for any bounded uniformly continuous function $\varphi : M \times N \to \mathbb{R}$. This means that $(F_k)_* m_k$ converges to $F_* m$ in the weak* topology. The hypothesis also implies that $(F_k)_* m_k = m_k$ converges to m in the weak* topology. It follows that $F_*m = m$, as we wanted to prove. Observe that

$$\int (\varphi \circ F_k)\, dm_k - \int (\varphi \circ F)\, dm_k \to 0 \quad \text{as } k \to \infty,$$

because $\varphi \circ F_k - \varphi \circ F$ converges uniformly to zero. Thus, to prove (5.13) we only have to show that

$$\int (\varphi \circ F)\, dm_k - \int (\varphi \circ F)\, dm \to 0 \quad \text{as } k \to \infty, \tag{5.14}$$

for any bounded uniformly continuous function $\varphi : M \times N \to \mathbb{R}$. This is clear if F is continuous, because m_k is assumed to converge to m in the weak* topology. This proves the second claim in the proposition. Now we prove the first claim, where F is not assumed to be continuous. Arguing as in Exercise 5.13, it is no restriction to suppose that $\varphi : M \times N \to \mathbb{R}$ is such that $\varphi(x, v)$ depends only on $(x_{-n}, \ldots, x_0, \ldots, x_{n-1}, v)$, for some fixed $m \geq 1$. For each $x \in M$, consider the continuous function $\Phi_x : N \to \mathbb{R}$ defined by

$$\Phi_x(v) = \varphi \circ F(x, v) = \varphi(f(x), F_x(v)).$$

Since Φ_x depends only on $(x_{-n+1}, \ldots, x_0, \ldots, x_n)$, we may view $\Phi : x \mapsto \Phi_x$ as a (measurable) map from X^{2n} to the set $C^0(N)$ of continuous functions on N. This is a separable metric space for the norm of uniform convergence. Thus, we may apply the theorem of Lusin (see [114, Appendix A.3]) to find, for any given $\varepsilon > 0$, a compact set $K \subset X^{2n}$ such that $p^{2n}(K^c) < \varepsilon$ and Φ is continuous on K. By construction, the restriction of $\varphi \circ F$ to $K \times N \to \mathbb{R}$ is continuous. Then, by the extension theorem of Tietze, there exists a continuous function $\psi : X^{2n} \times N \to \mathbb{R}$ such that $\sup|\psi| \leq \sup|\varphi|$ and $\psi(x, v) = \varphi \circ F(x, v)$ for every $x \in K$ and $v \in N$. Then (functions on $X^{2n} \times N$ are also viewed as functions on $M \times N$ that depend only on $(x_{-n+1}, \ldots, x_0, \ldots, x_n)$),

$$\begin{aligned}\int (\varphi \circ F)\, dm_k - \int (\varphi \circ F)\, dm &= \int \psi\, dm_k - \int \psi\, dm \\ &\quad + \int_{K^c \times N} (\varphi \circ F - \psi)\, dm_k - \int_{K^c \times N} (\varphi \circ F - \psi)\, dm.\end{aligned} \tag{5.15}$$

The first difference on the right-hand side is less than ε in absolute value if k is large, because ψ is continuous and m_k converges to m in the weak* topology. The remainder terms in (5.15) are bounded by

$$2\sup|\varphi| \left(m_k(K^c \times N) + m(K^c \times N)\right) = 2\sup|\varphi| \left(p_k^{2n}(K^c) + p^{2n}(K^c)\right).$$

The hypothesis that p_k converges to p implies that p_k^{2n} converges to p^{2n}, in the uniform sense (Exercise 5.14). In particular, $p_k^{2n}(K^c) < \varepsilon$ for all large k.

Substituting these estimates in (5.15) we find that

$$\left|\int(\varphi\circ F)\,dm_k - \int(\varphi\circ F)\,dm\right| < \varepsilon + 4\sup|\varphi|\,\varepsilon$$

if k is large enough. This completes the proof of (5.14) and the proposition. □

5.5 Disintegrations of s-states and u-states

For future use, we give an alternative characterization of s-states and u-states, in terms of their disintegrations along the vertical fibers. Take X, $M = X^{\mathbb{Z}}$ and N to be separable complete metric spaces.

5.5.1 Conditional probabilities

Let m be any probability measure on $M \times N$ that projects down to μ. A *disintegration* of m along vertical fibers is a measurable family $\{m_x : x \in M\}$ of probability measures on N satisfying

$$m(E) = \int_M m_x(\{v : (x,v) \in E\})\,d\mu(x) \quad \text{for any measurable set } E \subset M \times N.$$

The m_x are called *conditional probabilities* of m.

The family $\mathscr{P} = \{v : (x,v) \in E\}$ of vertical fibers is a measurable partition of $M \times N$, in the sense of Rokhlin; that is, it is the limit of some increasing sequence of finite partitions $\mathscr{P}_1 \prec \cdots \prec \mathscr{P}_n \prec \cdots$. To see this, just consider any countable basis of open sets $\{U_k : k \in \mathbb{N}\}$ of M, and take

$$\mathscr{P}_n = \bigvee_{k=1}^{n} \{U_k \times N, U_k^c \times N\}.$$

Then, Rokhlin's disintegration theorem (see [114, Section 5.2]) gives that a disintegration along vertical fibers does exist. Moreover, it is essentially unique: any two disintegrations coincide on a full μ-measure subset. Indeed, the proof of Rokhlin's theorem provides the following explicit characterization of the conditional probabilities:

$$m_x(B) = \lim_{U_k \to x} \frac{m(U_k \times B)}{m(U_k)} \quad \text{for } \mu\text{-almost every } x \in M \tag{5.16}$$

and any $B \subset N$ in some countable algebra $\mathscr{A}_N$ generating the Borel σ-algebra of N. The limit is taken over suitable decreasing sequences of open sets U_k whose intersection consists of the point x only.

Proposition 5.20 *An F-invariant probability measure m on $M \times N$ with $P_* m = \mu$ is an s-state if and only if it admits a disintegration that factors through M^+; that is, such that each m_x depends only on $x^+ = \pi^+(x)$. Dually, m is a u-state if and only if it admits a disintegration that factors through M^-.*

Proof Suppose that m admits a disintegration $\{m_x : x \in M\}$ where each m_x depends only on $x^+ = \pi^+(x)$: we write $m_x = m_{x^+}$. For any A^-, A^+, B,

$$\frac{m(A^- \times A^+ \times B)}{\mu(A^- \times A^+)} = \frac{\int_{A^-} d\mu^-(x^-) \int_{A^+} m_{x^+}(B)\, d\mu^+(x^+)}{\mu^-(A^-)\mu^+(A^+)}.$$

The right-hand side may be rewritten as

$$\frac{\int_{A^+} m_{x^+}(B)\, d\mu^+(x^+)}{\mu^+(A^+)} = \frac{m(M^- \times A^+ \times B)}{\mu(M^- \times A^+)},$$

and so it does not depend on A^-. Conversely, suppose m is a u-state. Fix countable bases of open sets $\{U_k^- : k \in \mathbb{N}\}$ of M^- and $\{U_l^+ : l \in \mathbb{N}\}$ of M^+. The products $U_k^- \times U_l^+$ form a basis of open sets of M. As observed in (5.16), there exists a full μ-measure set $M_0 \subset M$ and a countable algebra $\mathscr{A}_N$ generating the Borel σ-algebra of N, such that

$$m_x(B) = \lim_{U_k^- \times U_l^+ \to x} \frac{m(U_k^- \times U_l^+ \times B)}{\mu(U_k^- \times U_l^+)} \quad \text{for all } x \in M_0 \text{ and } B \in \mathscr{A}_N, \tag{5.17}$$

for every $x \in M_0$ and $B \in \mathscr{A}_N$, where the limit is taken over any sequence of basis elements $U_k^- \times U_l^+$ shrinking down to x. Given any pair $x, y \in M_0$ such that $\pi^+(x) = \pi^+(y)$, we may consider the same neighborhoods U_l^+ for both points. Then, as the right-hand side of (5.17) does not depend on U_k^-, we get that $m_x(B) = m_y(B)$ for every $B \in \mathscr{A}_N$. Since the algebra $\mathscr{A}_N$ is generating, this implies that $m_x = m_y$. We have shown that m_x factors through M^+, restricted to the full measure subset M_0. Then we can force this property on the whole M by redefining the m_x appropriately for $x \in M \setminus M_0$ (since this is a zero measure set, the family m_x continues to be a disintegration of m). This proves the claim for s-states. The dual claim, for u-states, is analogous. □

5.5.2 Martingale construction

Let $(X, \mathscr{B}, \mu)$ be a probability space. The *conditional expectation* $\mathbb{E}(\Psi \mid \mathscr{A})$ of an integrable function $\Psi : X \to \mathbb{R}$ relative to a σ-algebra $\mathscr{A} \subset \mathscr{B}$ is the unique $\mathscr{A}$-measurable function on X such that

$$\int_A \mathbb{E}(\Psi \mid \mathscr{A})\, d\mu = \int_A \Psi d\mu \quad \text{for all } A \in \mathscr{A}.$$

A *martingale* is a sequence $(\Psi_n, \mathscr{B}_n)$, $n \geq 0$ where $\mathscr{B}_n$, $n \geq 0$ is a non-decreasing sequence of sub-σ-algebras of $\mathscr{B}$, each $\Psi_n : X \to \mathbb{R}$, $n \geq 0$ is $\mathscr{B}_n$-measurable and integrable, and

$$\mathbb{E}(\Psi_{n+1} \mid \mathscr{B}_n) = \Psi_n \text{ almost everywhere, for every } n \geq 0. \tag{5.18}$$

Theorem 5.21 (martingale convergence theorem) *Given any martingale* $(\Psi_n, \mathscr{B}_n)$, $n \geq 0$, *such that* $\sup_n \mathbb{E}(|\Psi_n|) < \infty$, *there exists a measurable function* $\Psi : X \to \mathbb{R}$ *with* $\mathbb{E}(|\Psi|) < \infty$, *such that*

(1) $\Psi_n \to \Psi$ *almost everywhere;*
(2) $\mathbb{E}(\Psi \mid \mathscr{B}_n) = \Psi_n$ *almost everywhere, for every* $n \geq 0$;
(3) Ψ *is* $\mathscr{B}_\infty$-*measurable, where* $\mathscr{B}_\infty$ *denotes the* σ-*algebra generated by the union of all* $\mathscr{B}_n$, $n \geq 0$.

A proof of this classical theorem can be found in Durrett [49, § 5.2], for example. We are going to use the martingale convergence theorem to give a useful alternative proof of Proposition 5.17 in the case when is N a compact metric space.

Let $F : M \times N \to M \times N$ and $F^+ : M^+ \times N \to M^+ \times N$ be as in the previous section. Let m^+ be any F^+-invariant probability measure that projects down to μ^+ and let $\{m_y^+ : y \in M^+\}$ be a disintegration of m^+. Given any $x \in M$, define $m_x^+ = m_{x^+}^+$ where $x^+ = \pi^+(x)$.

Lemma 5.22 *The weak* limit* $m_x = \lim_n A^n(f^{-n}(x))_* m_{f^{-n}(x)}^+$ *exists for* μ-*almost every* $x \in M$.

Proof Given any continuous function $\varphi : N \to \mathbb{R}$ and $n \geq 0$, define

$$\Psi_{\varphi,n}(x) = \int \varphi \, d\big(A^n(f^{-n}(x))_* m_{f^{-n}(x)}^+\big). \tag{5.19}$$

For each $n \geq 0$, let $\mathscr{F}_n$ be the σ-algebra generated by the *cylinders*

$$[-n : \Delta_{-n}, \ldots, \Delta_l] = \{x \in M : x_i \in \Delta_i \text{ for } i = -n, \ldots, l\}$$

with $l \geq -n$ and $\Delta_i \subset X$ a measurable set for $i = -n, \ldots, l$. It is clear that $\mathscr{F}_n \subset \mathscr{F}_{n+1}$ for every $n \geq 0$. Since m_y^+ depends only on $y^+ = \pi^+(y)$ and $A(x)$ depends only on the zeroth coordinate x_0, the value of $\Psi_{\varphi,n}(x)$ depends only on x_k, $k \geq -n$. Hence, every $\Psi_{\varphi,n}$ is $\mathscr{F}_n$-measurable. Moreover, given any cylinder $C = [-n : \Delta_{-n}, \ldots, \Delta_l]$,

$$\begin{aligned}\int_C \Psi_{\varphi,n+1}(x) \, d\mu(x) &= \int_C \int \varphi \circ A^{n+1}(f^{-n-1}(x)) \, dm_{f^{-n-1}(x)}^+ \, d\mu(x) \\ &= \int_C \int \varphi \circ A^n(f^{-n}(x)) \, dq_{f^{-n-1}(x)} \, dp(x_{-n}) \cdots dp(x_{-1})\end{aligned}$$

where $q_y = \int_X A(y)_* m_y^+ \, dp(y_0)$. Observe that $dq_y = m_{f(y)}^+$ for μ-almost every y, because m^+ is F^+-invariant (Exercise 5.20). Therefore,

$$\int_C \Psi_{\varphi,n+1}(x)\, d\mu(x) = \int_C \int \varphi \circ A^n(f^{-n}(x))\, dm_{f^{-n}(x)}^+ \, d\mu(x)$$
$$= \int_C \Psi_{\varphi,n}(x)\, d\mu(x).$$

Since these cylinders C generate the σ-algebra $\mathscr{F}_n$, this proves that $(\Psi_{\varphi,n}, \mathscr{F}_n)$ is a martingale.

Then, by the martingale convergence theorem, there exists a measurable function $\Psi_\varphi : M \to \mathbb{R}$ such that

$$\Psi_\varphi(x) = \lim \Psi_{\varphi,n}(x) \tag{5.20}$$

for every x in some full μ-measure set $M_\varphi \subset M$. Now we use once more the fact that the space of continuous functions on the compact metric space N admits countable dense subsets. Taking the intersection of the M_φ over all φ in such a dense subset one obtains a full μ-measure set $M_* \subset M$ such that (5.20) holds for every $x \in M_*$ and every continuous function $\varphi : N \to \mathbb{R}$. Every map $\varphi \mapsto \Psi_\varphi(x)$ is linear and non-negative and it sends $\varphi \equiv 1$ to $\Psi_1(x) = 1$. Thus, by the Riesz representation theorem (see [114, Theorem A.3.11]), it defines a probability measure m_x on N:

$$\int \varphi \, dm_x = \Psi_\varphi(x), \quad \text{for each continuous } \varphi : N \to \mathbb{R}.$$

The definition gives that m_x is the weak* limit of $A^n(f^{-n}(x))_* m_{f^{-n}(x)}^+$. □

The corollary that follows means that the lift of m^+ is precisely the probability measure m on $M \times N$ that projects down to μ and admits the family $\{m_x : x \in M\}$ in Lemma 5.22 as a disintegration.

Corollary 5.23 *The lift of m^+ is the probability measure m defined on $M \times N$ by*

$$m(E) = \int m_x(E \cap (\{x\} \times N))\, d\mu(x)$$

for every measurable set $E \subset M \times N$.

Proof Recall the proof of Lemma 5.22. The definition (5.19) implies

$$m_{f(x)} = \lim_n A^{n+1}(f^{-n}(x))_* m_{f^{-n}(x)} = A(x)_* m_x$$

for μ-almost every x. Therefore (Exercise 5.20), the probability measure m is invariant under F. Moreover, given any cylinder $C = [0 : \Delta_0, \ldots, \Delta_l]$ in $\mathscr{F}_0$ and any continuous function $\varphi : N \to \mathbb{R}$,

$$\begin{aligned}\int \mathscr{X}_C \varphi \, dm &= \int_C \int \varphi \, dm_x \, d\mu(x) = \int_C \Psi_\varphi(x) \, d\mu(x) \\ &= \int_C \Psi_{\varphi,0}(x) \, d\mu(x) = \int_C \int \varphi \, dm_x^+ \, d\mu(x) = \int \mathscr{X}_C \varphi \, dm^+\end{aligned}$$

because $\mathbb{E}(\Psi_\varphi \mid \mathscr{F}_0) = \Psi_{\varphi,0}$. This proves that m projects to m^+ under the map $\Pi^+ : M \times N \to M^+ \times N$. □

Corollary 5.24 *Suppose that $m^+ = \mu^+ \times \eta$ for some forward stationary measure η. Then the lift m is a u-state.*

Proof Recall the proof of Lemma 5.22. In the present case, $m_x^+ \equiv \eta$ and so each

$$\Psi_{\varphi,n}(x) = \int \varphi \, d\big(A^n(f^{-n}(x))_* \eta\big)$$

depends only on $x_{-n}, \ldots, x_{-1}$. Thus, the limit $\Psi_\varphi(x) = \int \varphi \, dm_x$ depends only on $x^- = \pi^-(x)$. This means that m is a u-state. □

5.5.3 Remarks on 2-dimensional linear cocycles

Let $F : M \times \mathbb{R}^2 \to M \times \mathbb{R}^2$, $F(x,v) = (f(x), A(x)v)$ be an invertible locally constant linear cocycle and $\mathbb{P}F : M \times \mathbb{P}\mathbb{R}^2 \to M \times \mathbb{P}\mathbb{R}^2$ be the associated projective cocycle. By definition, the base dynamics $f : (M,\mu) \to (M,\mu)$ is a two-sided Bernoulli shift. The function $A : M \to \mathrm{GL}(2)$ is assumed to satisfy the integrability conditions $\log^+ \|A^{\pm 1}\| \in L^1(\mu)$. Take the Lyapunov exponents λ_+ and λ_- to be distinct and let $\mathbb{R}^2 = E_x^+ \oplus E_x^-$ be the Oseledets decomposition.

Let m^s and m^u be defined on the product space $M \times \mathbb{P}\mathbb{R}^2$ by

$$\begin{aligned}m^s(C \times D) &= \mu\big(\{x \in C : E_x^- \in D\}\big) = \int_C \delta_{E_x^-}(D) d\mu(x) \\ m^u(C \times D) &= \mu\big(\{x \in C : E_x^+ \in D\}\big) = \int_C \delta_{E_x^+}(D) d\mu(x)\end{aligned}$$

for all measurable sets $C \subset M$ and $D \subset \mathbb{P}\mathbb{R}^2$. In other words, m^s and m^u are the probability measures on $M \times \mathbb{P}\mathbb{R}^2$ that project down to μ and admit, respectively, $\delta_{E_x^-}$ and $\delta_{E_x^+}$ as conditional probabilities along the projective fibers. Note that m^s and m^u are $\mathbb{P}F$-invariant, because (Exercise 5.20)

$$A(x)_* \delta_{E_x^-} = \delta_{E_{f(x)}^-} \quad \text{and} \quad A(x)_* \delta_{E_x^+} = \delta_{E_{f(x)}^+} \quad \mu\text{-almost everywhere.}$$

Let $G \subset M \times \mathbb{P}\mathbb{R}^2$ be a measurable set invariant under $\mathbb{P}F$. Then the set $C^s = \{x \in M : (x, E_x^-) \in G\}$ is invariant under f, and so $\mu(C^s)$ is either 0 or 1.

Consequently, $m^s(G)$ is either 0 or 1, and this proves that m^s is ergodic. A similar argument proves that m^u is ergodic. Moreover, m^s is an s-state, because E^s_x depends only on $x^- = \pi^-(x)$; similarly, m^u is a u-state.

Lemma 5.25 *Any $\mathbb{P}F$-invariant probability measure m that projects down to μ may be written as a convex combination $m = am^s + bm^u$, with a, $b \geq 0$ and $a+b=1$*

Proof For each $k \geq 1$, consider the set

$$B_k = \left\{(x,v) \in M \times \mathbb{P}\mathbb{R}^2 : |\sin\measuredangle(v, E^*_x)| \geq \frac{1}{k}|\sin\measuredangle(E^s_x, E^u_x)| \text{ for } * = s,u\right\}.$$

Consider any $(x,v) \in B_k$. Since $\lambda_- < \lambda_+$, the angle between $A^n(x)v$ and $E^u_{f^n(x)}$ decays exponentially fast as $n \to +\infty$ (Exercise 3.5). On the other hand, by the theorem of Oseledets, the angle between $E^s_{f^n(x)}$ and $E^u_{f^n(x)}$ decays sub-exponentially. This implies that $F^n(x,v)$ eventually leaves B_k. So, by Poincaré recurrence, $m(B_k) = 0$ for every F-invariant probability measure m and every $k \geq 1$. This means that m is concentrated on the set of points $(x,v) \in M \times \mathbb{P}\mathbb{R}^2$ with $v \in \{E^s_x, E^u_x\}$, and so its disintegration has the form

$$m_x = a(x)\delta_{E^s_x} + b(x)\delta_{E^u(x)}$$

with $a(x), b(x) \geq 0$ and $a(x)+b(x) = 1$. Moreover, the functions $a(\cdot)$ and $b(\cdot)$ are (f,μ)-invariant and so, by ergodicity of μ, they are constant on a full μ-measure set. □

Remark 5.26 The hypotheses that F is locally constant and the system (f,μ) is a Bernoulli shift are not used in the proof of Lemma 5.25. Thus, the conclusion extends to any 2-dimensional linear cocycle over an ergodic system satisfying the integrability conditions in Theorem 3.20 and such that $\lambda_- < \lambda_+$. This will be used in the proof of Proposition 9.13.

Thus, m^s and m^u are the unique ergodic $\mathbb{P}F$-invariant probability measures. The next result also follows from the lemma.

Corollary 5.27 *Either m^s is the unique s-state or all $\mathbb{P}F$-invariant probability measures that project down to μ are s-states. Either m^u is the unique u-state or all $\mathbb{P}F$-invariant probability measures that project down to μ are u-states.*

Proof Suppose that there exists some s-state $m \neq m^s$. Then $m = am^s + bm^u$ with $b \neq 0$, and so we may write

$$m^u = \frac{1}{b}m - \frac{a}{b}m^s.$$

This implies that m^u is an s-state, and therefore so is any convex combination of m^s and m^u. By Lemma 5.25, this proves the first claim. The second one is analogous. □

In the next chapter, especially in Section 6.1, we will push this kind of picture to linear cocycles in arbitrary dimension.

5.6 Notes

The books of Kifer [72] and Arnold [5] contain systematic developments of the theory of random transformations, from two different perspectives. Our definition is more restrictive than Arnold's, as we consider only the Bernoulli case. However, several parts of the material we present here have been developed in much more generality.

The notions of s-state and u-state were introduced by Bonatti, Gomez-Mont and Viana [37] and were further developed in a series of papers by Avila and Viana [16], by Avila, Santamaria and Viana [13] and by Avila, Viana and Wilkinson [17]. The analysis of the invertible case presented in Section 5.4 is probably new.

The idea of disintegration into conditional probabilities relies on the disintegration theorem of Rokhlin [102]; see also [114, Section 5.2]. The martingale construction in Section 5.5.2 goes back to Furstenberg [55].

The remarks in Section 5.5.3 are from Avila and Viana [16]. They hold more generally: the cocycle need not be locally constant and it suffices that the invariant measure μ has *local product structure*, meaning that the restriction of μ to a neighborhood of every point x is equivalent to the product $\mu_x^+ \times \mu_x^-$ of a measure μ_x^+ on the unstable manifold of x by a measure μ_x^- on the stable manifold of x.

5.7 Exercises

Exercise 5.1 Show that if E is a measurable subset of $M \times N$ then:

(1) the sets $E_v = \{x \in M : (x,v) \in E\}$ and $E^x = \{v \in N : (x,v) \in E\}$ are measurable subsets of M and N, respectively, for every $v \in N$ and $x \in M$;
(2) the functions $N \to \mathbb{R}$, $v \mapsto \xi(E_v)$ and $M \to \mathbb{R}$, $x \mapsto \eta(E^x)$ are measurable, for any probability measures ξ on $(M, \mathscr{A})$ and η on $(N, \mathscr{B})$;
(3) moreover, $(\xi \times \eta)(E) = \int \xi(E_v)\, d\eta(v) = \int \eta(\xi^x)\, d\xi(x)$.

Exercise 5.2 Show that if $F : M \times N \to M \times N$, $F(x,v) = (f(x), F_x(v))$ is measurable then the maps $M \to N$, $x \mapsto F_x(v)$ and $F_x : N \to N$ are measurable, for every $v \in N$ and $x \in M$.

Exercise 5.3 Check that the transition operators can be defined in terms of the transition probabilities alone:

(1) $\mathscr{P}\varphi(v) = \int \varphi(w)\, dp(v,w)$ for every $v \in N$, and
(2) $\mathscr{P}^*\eta(B) = \int p(v,B)\, d\eta(v)$ for every measurable set $B \subset N$.

Exercise 5.4 Consider the situation in Example 5.1. Show that a probability measure η is stationary for the projective cocycle $\mathbb{P}F : M \times \mathbb{PR}^d \to M \times \mathbb{PR}^d$ if and only if it satisfies $\eta = \sum_{i=1}^m p_i (A_i)_*\eta$.

Exercise 5.5 Calculate all the stationary measures of the projective cocycle $\mathbb{P}F : X^{\mathbb{N}} \times \mathbb{PR}^2 \to X^{\mathbb{N}} \times \mathbb{PR}^2$ given by $\mathbb{P}F\big((\alpha_n)_{n\in\mathbb{N}}, [v]\big) = \big((\alpha_{n+1})_{n\in\mathbb{N}}, [\alpha_0 v]\big)$ where $X = \{A_1, A_2\}$ and $p = p_1\delta_{A_1} + p_2\delta_{A_2}$, with

$$A_1 = A_2^{-1} = \begin{pmatrix} \sigma & 0 \\ 0 & \sigma^{-1} \end{pmatrix} \quad \text{for some } \sigma > 1 \text{ and } p_1, p_2 > 0.$$

Exercise 5.6 Let η be a stationary measure and $\varphi : N \to \mathbb{R}$ be an η-stationary function. Show that there is a stationary function $\psi : N \to \mathbb{R}$ with $\varphi(v) = \psi(v)$ for η-almost every $v \in N$. Show that if φ is a characteristic function of some subset of N then ψ may also be taken to be a characteristic function.

Exercise 5.7 Let η be a stationary measure and $\varphi_n : N \to \mathbb{R}$, $n \geq 1$, be a monotone sequence of η-stationary functions converging η-almost everywhere to some bounded measurable function $\varphi : N \to \mathbb{R}$. Check that φ is η-stationary.

Exercise 5.8 Let $\phi : M \to \mathbb{R}$ be a bounded measurable function and $C \in \mathbb{R}$.

(1) Show that if $\{x \in M : \phi(x) > C\}$ has positive μ-measure then there exist $k, x_0, \ldots, x_{k-1}$ such that $\int \phi(x)\, d\mu(x_k, x_{k+1}, \ldots) > C$.
(2) Deduce that if $\int \phi(x)\, d\mu(x_k, x_{k+1}, \ldots) = C$ for every k, x_0, $\ldots$, x_{k-1} then $\phi(x) = C$ for μ-almost every $x \in M$.

Exercise 5.9 Check that the projection to N of a probability measure m invariant under a random transformation $F : M \times N \to M \times N$ need not be a stationary measure for F.

Exercise 5.10 Check that a stationary measure η is ergodic if and only if it is extremal; that is, if it cannot be written as a convex combination $\eta = \alpha_1\eta_1 + \alpha_2\eta_2$ where α_1, $\alpha_2 > 0$ with $\alpha_1 + \alpha_2 = 1$, and the probability measures η_1, η_2 are stationary and mutually singular (two measures η_1 and η_2 are mutually singular if there is a measurable subset such that $\eta_1(A) = 0$ and $\eta_2(A^c) = 0$).

Exercise 5.11 Let $F : M \times N \to M \times N$ be a one-sided random transformation and η be a stationary measure. Let $\int \xi \, d\hat{\eta}(\xi)$ be an ergodic decomposition of η. Show that $\int (\mu \times \xi) \, d\hat{\eta}(\xi)$ is an ergodic decomposition of the F-invariant probability measure $m = \mu \times \eta$.

Exercise 5.12 Let (X_i, d_i), $i \in I$, be separable, complete metric spaces, where $I \subset \mathbb{Z}$. Let X be the cartesian product $\Pi_{i \in I} X_i$, endowed with the product topology. Show that X is a separable space and

$$d(x,y) = \sum_{i \in I} 2^{-|i|} \min\{1, d_i(x_i, y_i)\}, \quad x, y \in X, \tag{5.21}$$

defines a complete metric, compatible with the product topology on X.

Exercise 5.13 Let (X_i, d_i), $i \in I$, be metric spaces, where $I \subset \mathbb{Z}$, and let the cartesian product $X = \prod_{i \in I} X_i$ be endowed with the distance (5.21). For any bounded uniformly continuous function $\varphi : X \to \mathbb{R}$ and any $\varepsilon > 0$, find a finite set $J \subset I$ and a bounded uniformly continuous function $\psi : X \to \mathbb{R}$ such that $\sup |\varphi - \psi| < \varepsilon$ and $\psi(x)$ depends only on the coordinates x_j with $j \in J$.

Exercise 5.14 Let (X_i, d_i), $i \in I$ be separable metric spaces, where $I \subset \mathbb{Z}$, and let the cartesian product $X = \prod_{i \in I} X_i$ be endowed with the distance (5.21). Prove that:

(1) If I is finite then the following is true relative to the uniform topology, the pointwise topology and the weak* topology: if $(p_{n,i})_n$ converges to p_i for every $i \in I$ then $(p_n = \prod_{i \in I} p_{n,i})_n$ converges to $p = \prod_{i \in I} p_i$. Indeed:
   (a) $\|p_n - p\| \leq \sum_{i \in I} \|p_{n,i} - p_i\|$;
   (b) $\{E \subset X : p_n(E) \to p(E)\}$ is a monotone class and contains every finite union of measurable rectangles $\prod_{i \in I} E_i$;
   (c) for any bounded uniformly continuous $\varphi : X \to \mathbb{R}$ and $\varepsilon > 0$ there is a countable partition $\{R_k : k \geq 1\}$ of X into rectangles with $p(\partial R_k) = 0$ and $|\varphi(z) - \varphi(w)| \leq \varepsilon$ if z and w are in the same R_k.

(2) When I is infinite, the statement remains true for the weak* topology (use Exercise 5.13) but, in general, not for the pointwise topology nor the uniform topology: Take $X_i = [0,1]$ and $p_{n,i} = (1 - n^{-1})\delta_0 + n^{-1}\delta_1$.

Exercise 5.15 Consider the situation in Example 5.1. Let η be a probability measure on N. Show that:

(1) η is forward stationary if and only if $\eta(B) = \sum_{i=1}^{m} p_i \eta(A_i^{-1}(B))$ for every measurable $B \subset N$;

(2) η is backward stationary if and only if $\eta(B) = \sum_{i=1}^{m} p_i \eta(A_i(B))$ for every measurable $B \subset N$.

The sets of backward and forward stationary measures may be disjoint.

Exercise 5.16 Let $\phi : M \to \mathbb{R}$ be a bounded measurable function and

$$\phi_k : M \to \mathbb{R}, \quad \phi_k(x) = \int \phi(x)\, d\mu(x_k, x_{k+1}, \ldots)$$

for each $k \geq 1$. Use the martingale convergence theorem to prove that $(\phi_k(x))_k$ converges to $\phi(x)$ for μ-almost every x.

Exercise 5.17 Prove that an F-invariant probability measure m on $M \times N$ is

(1) an s-state if and only if $m = \mu^- \times m^+$ for some probability measure m^+ on $M^+ \times N$; then $m^+ = \Pi_*^+ m$ and it is F^+-invariant;
(2) a u-state if and only if $m = \mu^+ \times m^-$ for some probability measure m^- on $M^- \times N$; then $m^- = \Pi_*^- m$ and it is F^--invariant;
(3) an su-state if and only if $m = \mu \times \eta$ where η, the projection of m to N, is both forward stationary and backward stationary.

Exercise 5.18 Check that the conditions (5.11) and (5.12) are preserved by, respectively, backward and forward iteration under the random transformation. Namely,

(1) if m is a probability measure on $M \times N$ that projects down to μ and satisfies (5.11), then the same is true for $F_*^{-1} m$;
(2) if m is a probability measure on $M \times N$ that projects down to μ and satisfies (5.12), then the same is true for $F_* m$.

Exercise 5.19 Prove the following.

(1) The space $\mathcal{M}(\mu)$ of probability measures on $M \times N$ that project down to μ is non-empty and compact for the weak* topology.
(2) The push-forward $F_* : \mathcal{M}(\mu) \to \mathcal{M}(\mu)$ is continuous.
(3) The subset of measures $m \in \mathcal{M}(\mu)$ satisfying (5.12) is invariant under F_* and is closed in the weak* topology.
(4) Any Cesaro limit of the forward iterates of any element of this subset is an F-invariant probability measure and a u-state for F.
(5) A dual construction yields s-states.

Exercise 5.20 Let m be a probability measure on $M \times N$ that projects down to an f-invariant probability measure μ. Prove that:

(1) if $f : M \to M$ is invertible (two-sided shift) then m is F-invariant if and only if $(F_x)_* m_x = m_{f(x)}$ for μ-almost every $x \in M$;

(2) if $f : M \to M$ is a one-sided shift then m is F-invariant if and only if

$$m_{f(x)} = \int (F_{y_0})_* m_{(y_0,x_1,\dots,x_n,\dots)}\, dp(y_0)$$

for μ-almost every $x = (x_0, x_1, \dots, x_n, \dots) \in M$.

Exercise 5.21 Let m be the lift of an F^+-invariant probability measure m^+. Show that m is an s-state if and only if any of the following conditions holds:

(1) $m = \mu^- \times m^+$;
(2) $m_x = m^+_{x^+}$ for μ-almost every $x \in M$, where $x^+ = \pi^+(x)$;
(3) $(F^+_{x^+})_* m_{x^+} = m^+_{f^+(x^+)}$ for μ^+-almost every $x^+ \in M^+$;

where $\{m_x : x \in M\}$ and $\{m^+_{x^+} : x^+ \in M^+\}$ are disintegrations of m and m^+.

Exercise 5.22 Prove that if m is an F-invariant probability measure on $M \times N$ that projects down to μ then its ergodic components also project down to μ.

Exercise 5.23 Prove that if m is an s-state (respectively, a u-state) then its ergodic components are s-states (respectively, u-states).

Exercise 5.24 Let $F : M \times \mathbb{R}^2 \to M \times \mathbb{R}^2$ be an invertible locally constant linear cocycle over a two-sided Bernoulli shift. Show that if F is pinching and twisting then the projective cocycle $\mathbb{P}F$ admits a unique u-state and a unique s-state, and they are distinct. Conclude that the forward stationary measure and the backward stationary measure are also unique and distinct.

6

Exponents and invariant measures

The main theme in the present chapter is that the Lyapunov exponents and Oseledets subspaces of a linear cocycle F may be retrieved from the invariant measures and the stationary measures of the associated projective cocycle $\mathbb{P}F$. This theme will be substantiated by three important results.

A theorem of Ledrappier [80], which we discuss in Section 6.1, states that the Lyapunov exponents of F coincide with the integrals of the function

$$\Phi : M \times \mathbb{P}\mathbb{R}^d \to \mathbb{R}, \quad \Phi(x,[v]) = \log \frac{\|A(x)v\|}{\|v\|} \tag{6.1}$$

with respect to the ergodic invariant probability measures of $\mathbb{P}F$. Moreover, any such probability measure is concentrated on the corresponding Oseledets sub-bundle.

When the cocycle is *strongly irreducible* and locally constant, the largest Lyapunov exponent is given by the integral of Φ with respect to *any* $\mathbb{P}F$-invariant measure of the form $\mu \times \eta$. That is the content of Furstenberg's formula [54], that we present in Section 6.2.

Finally, Section 6.3 is devoted to Furstenberg's [54] criterion for the extremal Lyapunov exponents of locally constant $\mathrm{SL}(2)$-dimensional cocycles to coincide: if $\lambda_- = \lambda_+$ then either the cocycle lives in a compact subgroup or it leaves invariant some finite subset of $\mathbb{P}\mathbb{R}^2$.

Throughout, $\mathbb{P}F : M \times \mathbb{P}\mathbb{R}^d \to M \times \mathbb{P}\mathbb{R}^d$ is the projective cocycle associated with a linear cocycle $F : M \times \mathbb{R}^d \to M \times \mathbb{R}^d$ defined over an ergodic transformation $f : (M,\mu) \to (M,\mu)$ by a measurable function $A : M \to \mathrm{GL}(d)$. We always take M to be a separable complete metric space and μ to be a Borel measure on M such that $\log^+ \|A^{\pm 1}\| \in L^1(\mu)$. The extremal Lyapunov exponents are denoted, indifferently, as $\lambda_\pm(F,\mu)$ or $\lambda_\pm(A,\mu)$.

For some of the results we take (f,μ) to be a Bernoulli shift and A to be locally constant; that is, such that $A(x)$ depends only on the zeroth coordinate of $x \in M$. Then F and $\mathbb{P}F$ are random transformations, in the sense of the previous chapter.

6.1 Representation of Lyapunov exponents

Let $\lambda_1 > \cdots > \lambda_k$ be the Lyapunov exponents of the linear cocycle F (keep in mind that we take (f, μ) to be ergodic) and then let $\mathbb{R}^d = V_x^1 \supset \cdots \supset V_x^k \supset \{0\}$ be the Oseledets flag of F (Theorem 4.1). When the cocycle is invertible, $\mathbb{R}^d = E_x^1 \oplus \cdots \oplus E_x^k$ denotes the Oseledets decomposition of F (Theorem 4.2).

Theorem 6.1 (Ledrappier) *Given any $\mathbb{P}F$-invariant ergodic probability measure m on $M \times \mathbb{P}\mathbb{R}^d$ that projects down to μ, there exists $j \in \{1, \ldots, k\}$ such that*

$$\int \Phi \, dm = \lambda_j \quad \textit{and} \quad m(\{(x,[v]) : v \in V_x^j \setminus V_x^{j+1}\}) = 1. \tag{6.2}$$

Conversely, given $j \in \{1, \ldots, k\}$ there is an $\mathbb{P}F$-invariant ergodic probability measure m projecting to μ and satisfying (6.2). *When F is invertible, one may replace $V_x^j \setminus V_x^{j+1}$ by E_x^j in* (6.2).

Proof Let m be $\mathbb{P}F$-invariant and ergodic. On the one hand, by ergodicity,

$$\lim_n \frac{1}{n} \log \frac{\|A^n(x)v\|}{\|v\|} = \lim_n \frac{1}{n} \sum_{i=0}^{n-1} \Phi \circ \mathbb{P}F^i(x,[v]) = \int \Phi \, dm \tag{6.3}$$

for m-almost every $(x,[v])$. On the other hand, for μ-almost every $x \in M$ and every $[v] \in \mathbb{P}\mathbb{R}^d$, the expression on left-hand side of (6.3) converges to some Lyapunov exponent λ_j. It follows that $\int \Phi \, dm = \lambda_j$ and m gives full weight to the set

$$\{(x,[v]) \in M \times \mathbb{P}\mathbb{R}^d : v \in V_x^j \setminus V_x^{j+1}\}$$

of pairs $(x,[v])$ for which the limit in (6.3) is λ_j. When F is invertible, one can take both limits $n \to \pm\infty$ in (6.3). We conclude that m gives full weight to the set

$$\{(x,[v]) \in M \times \mathbb{P}\mathbb{R}^d : v \in E_x^j\}$$

of pairs $(x,[v])$ for which the limit in (6.3) is λ_j for both $n \to +\infty$ and $n \to -\infty$.

To prove the converse statement we need a few auxiliary lemmas. Let $\mathscr{M}(\mu)$ denote the space of probability measures on $M \times \mathbb{P}\mathbb{R}^d$ that project down to μ.

Lemma 6.2 *The push-forward $F_* : \mathscr{M}(\mu) \to \mathscr{M}(\mu)$ is well defined and continuous relative to the weak* topology.*

Proof Let $(m_n)_n$ be a sequence in $\mathcal{M}(\mu)$ converging, in the weak* sense, to some probability measure m. We want to show that F_*m_n converges to F_*m; that is,

$$\int (\varphi \circ F)\, dm_n \to \int (\varphi \circ F)\, dm \tag{6.4}$$

for every bounded continuous function $\varphi : M \times \mathbb{P}\mathbb{R}^d \to \mathbb{R}$. Since $f : M \to M$ and $A : M \to \mathrm{SL}(d)$ are measurable, we can use Lusin's theorem to find, for every $\varepsilon > 0$, compact sets $K \subset M$ such that $\mu(K^c) < \varepsilon$ and the restriction of F to $L = K \times \mathbb{P}\mathbb{R}^d$ is continuous. Then, $\varphi \circ F$ is continuous restricted to L and, by the Tietze extension theorem, there is some continuous function $\psi : M \times \mathbb{P}\mathbb{R}^d \to \mathbb{R}$ such that $\sup|\psi| \le \sup|\varphi|$ and $\psi = \varphi \circ F$ restricted to L. It follows that

$$\Big|\int (\varphi \circ F)\, dm_n - \int (\varphi \circ F)\, dm\Big| \le \Big|\int \psi\, dm_n - \int \psi\, dm\Big| +$$
$$+ \Big|\int_{L^c} \psi\, dm_n - \int_{L^c} \psi\, dm\Big| + \Big|\int_{L^c} (\varphi \circ F)\, dm_n - \int_{L^c} (\varphi \circ F)\, dm\Big|$$

The first term on the right-hand side is smaller than ε if n is large enough, because $m_n \to m$ and ψ is continuous. The remain terms are bounded by

$$2 \sup|\varphi| \left(m_n(L^c) + m(L^c)\right) = 4 \sup|\varphi|\, \mu(K^c) \le 4 \sup|\varphi|\, \varepsilon.$$

This proves (6.4), and so the argument is complete. □

Lemma 6.3 *$\mathcal{M}(\mu)$ is sequentially compact, relative to the weak* topology.*

Proof We are going to use the fact that every Borel measure in a separable, completely metrizable space is tight (see [114, Appendix A.3]). This means that for any $\varepsilon > 0$ we may find a compact set $K \subset M$ such that $\mu(K) > 1 - \varepsilon$. Then $K \times \mathbb{P}\mathbb{R}^d$ is compact and $m(K \times \mathbb{P}\mathbb{R}^d) = \mu(K) > 1 - \varepsilon$ for every $m \in \mathcal{M}(\mu)$. This proves that the set $\mathcal{M}(\mu)$ is tight. Hence, by the theorem of Prohorov (see [114, Theorem 2.1.8]), it is sequentially compact. □

Lemma 6.4 *Let $x \mapsto V_x$ be a measurable sub-bundle of $M \times \mathbb{R}^d$. Then the subset of probability measures $m \in \mathcal{M}(\mu)$ such that $m(\{(x,[v]) : v \in V_x\}) = 1$ is closed in the weak* topology.*

Proof Let $(m_n)_n$ be a sequence in $\mathcal{M}(\mu)$ such that

$$m_n(\{(x,[v]) : v \in V_x\}) = 1 \quad \text{for all } n,$$

and which converges to some m in the weak* topology. We want to show that $m(\{(x,[v]) : v \in V_x\}) = 1$. Since the sub-bundle V is assumed to be measurable, for every $\varepsilon > 0$ one can use Lusin's theorem to find a compact set $K \subset M$ with $\mu(K) > 1 - \varepsilon$ and such that the restriction of V to K is continuous. Then

$\{(x,[v]) \in K \times \mathbb{P}\mathbb{R}^d : v \in V_x\}$ is closed and so its m-measure is bounded below by

$$\limsup_n m_n(\{(x,[v]) \in K \times \mathbb{P}\mathbb{R}^d : v \in V_x\}) \geq \mu(K) > 1 - \varepsilon.$$

Since ε is arbitrary, this proves that $m(\{(x,[v]) : v \in V_x)\}) = 1$. □

We are now ready to conclude the proof of Theorem 6.1. Let j be fixed. Let us start with the invertible case. Since the Oseledets sub-bundle $x \mapsto E_x^j$ is measurable it follows from Exercise 4.1 that it admits a *measurable section*; that is, there exists a measurable vector field $x \mapsto \sigma(x)$ such that $\sigma(x) \in E_x^j$ for every x. Let m_0 be the probability measure in $\mathcal{M}(\mu)$ that admits $\delta_{\sigma(x)}$ as a disintegration. In other words,

$$m_0(B) = \mu\big(\{x \in M : (x, \sigma(x)) \in B\}\big)$$

for every measurable $B \subset M \times \mathbb{P}\mathbb{R}^d$. Then define, for $n \geq 1$,

$$m_n = \frac{1}{n} \sum_{i=0}^{n-1} (\mathbb{P}F)_*^i m_0.$$

It is clear that $m_n(\{(x,[v]) : v \in E_x^j\}) = 1$ for every $n \geq 0$. Then, by Lemma 6.4, the same is true for every accumulation point of the sequence $(m_n)_n$. Using Lemma 6.2, every accumulation point is a $\mathbb{P}F$-invariant probability measure. By Lemma 6.3, accumulation points do exist. Considering ergodic components, one gets that there exists some $\mathbb{P}F$-invariant ergodic probability measure m such that $m(\{(x,[v]) : v \in E_x^j\}) = 1$. Then $\tilde{\Phi}(x,[v]) = \lambda_j$ for m-almost every $(x,[v])$, and so $\int \Phi \, dm = \int \tilde{\Phi} \, dm = \lambda_j$. This finishes the proof in the invertible case.

The general case can be deduced easily, through the following invertible extension of $F : M \times \mathbb{R}^d \to M \times \mathbb{R}^d$. Let (see [114, Section 2.4.2])

$$\hat{f} : \hat{M} \to \hat{M}, \quad \hat{f}(\ldots, x_n, \ldots, x_{-1}, x_0) \mapsto (\ldots, x_n, \ldots, x_{-1}, x_0, f(x_0))$$

be the *natural extension* of $f : M \to M$, defined on the *space of pre-orbits*

$$\hat{M} = \{(x_n)_{n \leq 0} : f(x_n) = x_{n+1} \text{ for every } n < 0\}.$$

Denote by $\pi : \hat{M} \to M$ the projection to the zeroth coordinate. There is a unique $\hat{f}$-invariant measure on $\hat{M}$ such that $\pi_* \hat{\mu} = \mu$. Define $\hat{A} = A \circ \pi : \hat{M} \to \mathrm{GL}(d)$ and then let $\hat{F} : \hat{M} \times \mathbb{R}^d \to \hat{M} \times \mathbb{R}^d$ be the linear cocycle defined by $\hat{A}$ over $\hat{f}$. Note that

$$\int \log \|\hat{A}^{\pm 1}\| \, d\hat{\mu} = \int \log \|A^{\pm 1}\| \, d\mu.$$

It is also clear that F and $\hat{F}$ have the same Lyapunov exponents, and their Oseledets flags $(V^i)_i$ and $(\hat{V}^i)_i$ are related by

$$\hat{V}^i_{\hat{x}} = V^i_{\pi(\hat{x})} \quad \text{for every } \hat{x} \in \hat{M} \text{ and every } i.$$

By the previous paragraph, there exists some $\mathbb{P}\hat{F}$-invariant probability measure $\hat{m}$ on $\hat{M} \times \mathbb{R}^d$ that projects down to $\hat{\mu}$ and satisfies $\hat{m}(\{(x,[v]) : v \in \hat{E}^j_x\}) = 1$. Let m' be the image of $\hat{m}$ under the map $\pi \times \mathrm{id} : \hat{M} \times \mathbb{P}\mathbb{R}^d \to M \times \mathbb{P}\mathbb{R}^d$. Then m' is a $\mathbb{P}F$-invariant probability measure and

$$m'(\{(x,[v]) : v \in V^j_x \setminus V^{j+1}_x\}) = 1.$$

because $\hat{E}^j_{\hat{x}} \subset V^j_{\pi(\hat{x})} \setminus V^{j+1}_{\pi(\hat{x})}$ for every $\hat{x} \in \hat{M}$. Considering ergodic components, one concludes that there exists some $\mathbb{P}F$-invariant ergodic probability measure m such that $m(\{(x,[v]) : v \in V^j_x \setminus V^{j+1}_x\}) = 1$. The other claim in (6.2) follows immediately, just as in the invertible case. □

The following application of Theorem 6.1 will be useful later on. Take the linear cocycle $F : M \times \mathbb{R}^d \to M \times \mathbb{R}^d$ to be locally constant and, for the time being, invertible.

Proposition 6.5 *For invertible locally constant cocycles,*

(1) $\lambda_+(F,\mu) = \max\{\int \Phi\, dm : m \text{ a } u\text{-state for } \mathbb{P}F\}$,
(2) $\lambda_-(F,\mu) = \min\{\int \Phi\, dm : m \text{ an } s\text{-state for } \mathbb{P}F\}$.

Proof It follows from Theorem 6.1 and Exercise 5.22 that

$$\lambda_-(F,\mu) \le \int \Phi\, dm \le \lambda_+(F,\mu) \tag{6.5}$$

for every $\mathbb{P}F$-invariant probability measure m that projects down to μ. Thus, we need to show that the maximum is realized by some u-state, and the minimum is realized by some s-state.

Consider the Oseledets sub-bundle E^1 corresponding to the largest Lyapunov exponent $\lambda_1 = \lambda_+(F,\mu)$. As observed in Remark 4.3, the subspace E^1_x depends only on the negative part $x^- = \pi^-(x)$ of the point x. So we may find a measurable section $x \mapsto \sigma(x)$ to the sub-bundle E^1 such that $\sigma(x)$ depends only on x^-. Let m_0 be the probability measure in $\mathcal{M}(\mu)$ that admits $\{\delta_{\sigma(x)} : x \in M\}$ as a disintegration. In other words,

$$m_0(B) = \mu(\{x \in M : (x,\sigma(x)) \in B\})$$

for every measurable $B \subset M \times \mathbb{P}\mathbb{R}^d$. Then define, for $n \ge 1$,

$$m_n = \frac{1}{n}\sum_{j=0}^{n-1} (\mathbb{P}F)^j_* m_0.$$

Using Lemma 6.2 one easily gets that every accumulation point of this sequence is an $\mathbb{P}F$-invariant probability measure that projects down to μ. Accumulation points do exist, by Lemma 6.3. By Exercise 5.18, every m_n satisfies the u-state condition (5.12). Hence, according to Lemma 5.19, every accumulation point m is a u-state for $\mathbb{P}F$. Note also that $m_n(\{(x,[v]) : v \in E_x^j\}) = 1$ for every $n \geq 0$. Then, in view of Lemma 6.4, the same is true for every accumulation point m. This implies that

$$\lim_n \frac{1}{n} \sum_{j=0}^{n-1} \Phi(F^j(x,[v])) = \lambda_1$$

for m-almost every $(x,[v])$, and so $\int \Phi\, dm = \lambda_1 = \lambda_+(F,\mu)$. This shows that m realizes the maximum in (6.5), as required. A dual argument shows that the minimum in (6.5) is realized by some s-state. □

Now consider the non-invertible cocycles $F^\pm$ and $\mathbb{P}F^\pm$ associated with F and $\mathbb{P}F$, as introduced in Section 5.4. We have seen that every $\mathbb{P}F$-invariant probability measure m is the lift of some $\mathbb{P}F^+$-invariant probability measure m^+. Note that

$$\int_{M\times\mathbb{P}\mathbb{R}^d} \Phi\, dm = \int_{M^+\times\mathbb{P}\mathbb{R}^d} \Phi\, dm^+$$

because Φ depends only on the zeroth coordinate. Moreover, m is a u-state if and only if $m^+ = \mu^+ \times \eta$ for some forward stationary measure η. In this case,

$$\int_{M\times\mathbb{P}\mathbb{R}^d} \Phi\, dm = \int_{M^+\times\mathbb{P}\mathbb{R}^d} \Phi\, d(\mu^+ \times \eta) = \int_{M\times\mathbb{P}\mathbb{R}^d} \Phi\, d(\mu\times\eta).$$

Similar remarks apply to $\mathbb{P}F^-$-invariant probabilities and backward stationary measures. Thus, Proposition 6.5 immediately gives

Corollary 6.6 *For invertible locally constant cocycles,*

(a) $\lambda_+(F,\mu) = \max\{\int \Phi\, d(\mu\times\eta) : \eta$ *forward stationary for* $\mathbb{P}F\}$,
(b) $\lambda_-(F,\mu) = \min\{\int \Phi\, d(\mu\times\eta) : \eta$ *backward stationary for* $\mathbb{P}F\}$

This also leads to the one-sided version of Proposition 6.5:

Proposition 6.7 *For general locally constant cocycles,*

$$\lambda_+(F,\mu) = \max\left\{\int \Phi\, d(\mu\times\eta) : \eta \text{ a stationary measure for } \mathbb{P}F\right\}.$$

Proof This follows from the previous corollary applied to the invertible extension $\hat{F} : \hat{M}\times\mathbb{R}^d \to \hat{M}\times\mathbb{R}^d$ of the linear cocycle $F : M\times\mathbb{R}^d \to M\times\mathbb{R}^d$. More precisely, let $\hat{f} : \hat{M} \to \hat{M}$ be the two-sided shift on $\hat{M} = X^{\mathbb{Z}}$ and let $\hat{A} : \hat{M} \to \mathrm{GL}(d)$ be given by $\hat{A}(\hat{x}) = A(\pi(\hat{x}))$, where $\pi : \hat{M} \to M$ is the canonical projection. Define $\hat{F}(\hat{x},v) = (\hat{f}(\hat{x}),\hat{A}(\hat{x})v)$. It is clear that $\lambda_+(F,\mu) = \lambda_+(\hat{F},\hat{\mu})$, where $\hat{\mu} = p^{\mathbb{Z}}$. Moreover, a probability measure η on $\mathbb{P}\mathbb{R}^d$ is stationary for $\mathbb{P}F$ if and only if it is stationary for $\mathbb{P}\hat{F}$. □

6.2 Furstenberg's formula

In this section we take the linear cocycle $F : M \times \mathbb{R}^d \to M \times \mathbb{R}^d$ to be locally constant and the shift $f : (M,\mu) \to (M,\mu)$ to be one-sided.

6.2.1 Irreducible cocycles

A linear cocycle F is *irreducible* if there is no proper subspace of $\mathbb{R}^d$ invariant under $A(x)$ for μ-almost every $x \in M$; and it is *strongly irreducible* if there is no finite family of proper subspaces invariant by $A(x)$ for μ-almost every x.

Proposition 6.7 implies that the integral of the function (6.1) with respect to some $\mathbb{P}F$-invariant probability measure of the form $m = \mu \times \eta$ is equal to the largest Lyapunov exponent. We are going to see that if one assumes strong irreducibility then the same is true for *every* invariant probability measure of the form $\mu \times \eta$. That also implies that for $j > 1$ the measure m in Theorem 6.1 cannot be a product measure $\mu \times \eta$.

Theorem 6.8 (Furstenberg's formula) *If $F : M \times \mathbb{R}^d \to M \times \mathbb{R}^d$ is strongly irreducible then*

$$\lambda_+(F,\mu) = \int \Phi d(\mu \times \eta)$$

for any stationary measure η of the associated projective cocycle $\mathbb{P}F$.

Proof Let $m = \mu \times \eta$. By the ergodic theorem, the Birkhoff average

$$\Psi(x,[v]) = \lim_n \frac{1}{n} \log \frac{\|A^n(x)v\|}{\|v\|} = \lim_n \frac{1}{n} \sum_{i=0}^{n-1} \Phi \circ \mathbb{P}F^i(x,[v])$$

exists at m-almost every point, and it satisfies $\int \Psi\, dm = \int \Phi\, dm$. By the theorem of Oseledets, there exists $M_0 \subset M$ such that $\mu(M_0) = 1$ and

$$\lim_n \frac{1}{n} \log \frac{\|A^n(x)v\|}{\|v\|} = \lambda_1 \quad \text{for every } x \in M_0 \text{ and } v \notin V_x^2.$$

So, it suffices to check that the set of pairs (x,v) with $v \in V_x^2$ has zero m-measure. For this we need the lemma that follows. A measure ξ in a measurable space $(X,\mathscr{B})$ is *non-atomic* if $\xi(\{x\}) = 0$ for every $x \in X$.

Lemma 6.9 *If the cocycle F is strongly irreducible then $\eta(V) = 0$ for any proper projective subspace V of $\mathbb{PR}^d$ and any $\mathbb{P}F$-stationary measure η. In particular, every F-stationary measure η is non-atomic.*

Proof Suppose that there is some proper projective subspace V with $\eta(V) > 0$. Let $d_0 \geq 1$ be the smallest dimension of such subspaces, c be the maximum value of $\eta(V)$ over all subspaces of dimension d_0, and $\mathscr{V}$ be the family of all subspaces V of dimension d_0 such that $\eta(V) = c$. Since subspaces are closed subsets of $\mathbb{PR}^d$, we have $\eta(V) = c$ for any accumulation point V of any sequence $(V_n)_n$ of subspaces of dimension d_0 such that $\eta(V_n) \to c$. By compactness of the Grassmannian manifold $G(d_0, d)$, accumulation points do exist, and so the family $\mathscr{V}$ is non-empty. Moreover, $\mathscr{V}$ is finite: since η is a probability measure and the elements of $\mathscr{V}$ are essentially disjoint (due to the choice of d_0) we have that $\#\mathscr{V} \leq 1/c$. We write $\mathscr{V} = \{V_1, \ldots, V_n\}$. Since

$$c = \eta(V_i) = \int \eta\left(A(x)^{-1}(V_i)\right) d\mu(x)$$

and $\eta(A(x)^{-1}(V_i)) \leq c$ for all $x \in M$, we must have $\eta(A(x)^{-1}(V_i)) = c$ for μ-almost every $A(x)$. In view of the choice of $\mathscr{V}$, this implies that

$$A(x)^{-1}(V_i) \in \mathscr{V}$$

for μ-almost every $x \in M$. This means that the set $\mathscr{V}$ is invariant under almost every $A(x_0)$, which contradicts the strong irreducibility hypothesis. This contradiction proves the first part of the lemma. The last part is a special case (subspaces of dimension 1). □

Hence, $\eta(V_x^2) = 0$ for every $x \in M$ and so $m\big(\{(x,[v]) : v \in V_x^2\}\big) = 0$. It follows that $\int \Phi\, dm = \int \Psi\, dm = \lambda_1$ for any $\mathbb{P}F$-invariant probability measure of the form $m = \mu \times \eta$, as we wanted to prove. □

6.2.2 Continuity of exponents for irreducible cocycles

Let $\mathscr{I}$ be the space of pairs (A, p) where $A : X \to \mathrm{SL}(d)$ is a measurable function and p is a probability measure on X such that $\log^+ \|A^{\pm 1}\| \in L^1(p)$ and A is strongly irreducible: no finite family of subspaces of $\mathbb{R}^d$ is invariant under $A(x)$ for p-almost every x. For any $(A, p) \in \mathscr{I}$ we denote $\mu = p^{\mathbb{N}}$ and $\nu = A_* p$. Sometimes, we let ourselves view A as a function $M \to \mathrm{SL}(d)$ depending only on the zeroth coordinate. Then $\nu = A_* \mu$.

A sequence of measures $(\xi_k)_k$ in $\mathrm{GL}(d)$ is called *uniformly integrable* if given $\varepsilon > 0$, there exist $R > 0$ and $k_0 \geq 1$ such that

$$\int_{\{\|B\| > R\}} \log^+ \|B^{\pm 1}\|\, d\xi_k(B) < \varepsilon \tag{6.6}$$

for all $k \geq k_0$ and for both choices of the sign. This is the case, for example, if there is some compact set $K \subset \mathrm{GL}(d)$ such that $\xi_k(K) = 1$ for every k. Theorem 6.8 has the following interesting consequence:

Corollary 6.10 *Let $(A, p) \in \mathscr{I}$ and (A_k, p_k), $k \geq 1$ be a sequence in $\mathscr{I}$ such that the probability measures $\nu_k = (A_k)_* p_k$ converge to $\nu = A_* p$ in the weak* topology and are uniformly integrable. Then $\lambda_+(A_k, \mu_k)$, $k \geq 1$ converges to $\lambda_+(A, \mu)$ as $k \to \infty$, where $\mu_k = p_k^{\mathbb{N}}$.*

Proof For each $k \geq 1$, define $F_k(x, v) = (f(x), A_k(x_0)v)$ and then let η_k be any F_k-stationary measure. By Theorem 6.8,

$$\lambda_+(A_k, \mu_k) = \int \Phi(B, [v])\, d\nu_k(B)\, d\eta_k([v]), \tag{6.7}$$

where $\Phi(B, [v]) = \log\left(\|B(v)\| / \|v\|\right)$. Up to restricting to a subsequence of an arbitrary subsequence, we may suppose that η_k converges to some probability measure η on $\mathbb{PR}^d$. Note that η is F-stationary, by Proposition 5.9. Thus, Theorem 6.8 also gives

$$\lambda_+(A, \mu) = \int \Phi(B, [v])\, d\nu(B) d\eta([v]). \tag{6.8}$$

So, we only have to show that the right-hand side of (6.7) converges to the right-hand side of (6.8) when $k \to \infty$. Let $\varepsilon > 0$. Since $\log^+ \|A^{\pm 1}\| \in L^1(p)$, we may find $R > 0$ such that

$$\int_{\{\|B\| > R\}} \log^+ \|B^{\pm 1}\|\, d\nu(B) = \int_{\{\|A(x)\| > R\}} \log^+ \|A(x)^{\pm 1}\|\, dp(x) < \varepsilon.$$

The assumption of uniform integrability ensures that, increasing R if necessary, the same holds for ν_k if k is sufficiently large. Now observe that

$$|\Phi(B, [v])| \leq \log^+ \|B\| + \log^+ \|B^{-1}\| \quad \text{for every } (B, [v]) \in \mathrm{GL}(d) \times \mathbb{PR}^d.$$

Thus,

$$\left| \int_{\{\|B\| > R\}} \Phi\, d\nu_k\, d\eta_k - \int_{\{\|B\| > R\}} \Phi\, d\nu\, d\eta \right| \leq 4\varepsilon \tag{6.9}$$

for every large k. Next, observe that $\nu_k \times \eta_k$ converges to $\nu \times \eta$ in the weak* topology (Exercise 5.14). Since the function Φ is continuous, using Exercise 6.1 we get that

$$\left| \int_{\{\|B\| \leq R\}} \Phi\, d\nu_k\, d\eta_k - \int_{\{\|B\| \leq R\}} \Phi\, d\nu\, d\eta \right| \leq \varepsilon \tag{6.10}$$

for every large k. Combining (6.7)–(6.10) we get that $|\lambda_+(A_k, \mu_k) - \lambda(A, \mu)| < 5\varepsilon$ for every k sufficiently large. □

6.3 Theorem of Furstenberg

Let $F : M \times \mathbb{R}^2 \to M \times \mathbb{R}^2$ be a locally constant linear cocycle defined by a measurable function $A : X \to \mathrm{SL}(2)$ over a one-sided shift $f : M \to M$. We continue to denote $M = x^{\mathbb{N}}$ and $\mu = p^{\mathbb{N}}$ and $\nu = A_* p$.

The result that follows was stated before, in Theorem 1.2, but this time the hypotheses are formulated directly in terms of the support of the measure ν, rather than the monoid $\mathscr{B}$ generated by the support. Exercises 6.6 and 6.7 show that the two formulations are equivalent. In the next chapter we will discuss extensions of the statement to arbitrary dimensions.

Theorem 6.11 (Furstenberg) *Assume that*

(a) *the support of* ν *is not contained in a compact subgroup of* $\mathrm{SL}(2)$ *and*
(b) *there is no non-empty finite set* $L \subset \mathbb{P}\mathbb{R}^2$ *such that* $B(L) = L$ *for every* B *in the support of* ν.

Then the extremal Lyapunov exponents are distinct: $\lambda_- < 0 < \lambda_+$.

Recall that the *support* of a measure θ on a topological space is the (closed) subset $\operatorname{supp}\theta$ of points whose neighborhoods all have positive measure. It follows from the theorem (and Exercises 6.6 and 6.7) that there are four possibilities for $\mathrm{SL}(2)$-cocycles:

(1) the cocycle is non-pinching: $\operatorname{supp}\nu$ is contained in some compact subgroup of $\mathrm{SL}(2)$;
(2) the cocycle is non-twisting, with either one or two invariant subspaces: $\operatorname{supp}\nu$ is contained in a triangular subgroup or a diagonal subgroup;
(3) the cocycle is non-twisting, with an invariant set formed by two subspaces which are interchanged by some of the matrices;
(4) in all other cases, the Lyapunov exponents are distinct and non-zero.

This discussion is much refined by the following theorem of Thieullen [111], which classifies $\mathrm{SL}(2)$-cocycles up to conjugacy. A function $\phi : M \to \mathbb{R}$ is *cohomologous to zero* $\mod r$, for a given $r > 0$, if there exists a measurable function $u : M \to \mathbb{R}$ such that

$$\phi + u \circ f - u \in r\mathbb{Z} \quad \text{almost everywhere.} \tag{6.11}$$

Given a set $E \subset M$, we define $R_E : M \to \mathrm{SL}(2)$ by

$$R_E(x) = \begin{cases} \text{rotation of } \pi/2 & \text{if } x \in E \\ \text{id} & \text{otherwise.} \end{cases}$$

Theorem 6.12 (Thieullen) *Let (f,μ) be ergodic. If $A : M \to \mathrm{SL}(2)$ satisfies $\log^+ \|A^{\pm 1}\| \in L^1(\mu)$ then there exists $P : M \to \mathrm{SL}(2)$ such that the function $B(x) = P^{-1}(f(x)) \cdot A(x) \cdot P(x)$ has one of the following forms, μ-almost everywhere:*

(a) $B(x) = \begin{pmatrix} \cos\theta(x) & -\sin\theta(x) \\ \sin\theta(x) & \cos\theta(x) \end{pmatrix}$ *with θ not cohomologous to zero* $\mod \pi$.

(b) $B(x) = \begin{pmatrix} a(x) & b(x) \\ 0 & 1/a(x) \end{pmatrix}$ *with $\int \log|a|\, d\mu = 0$ and $\log|b| \in L^1(\mu)$.*

(c) $B(x) = R_E(x) \begin{pmatrix} cca(x) & 0 \\ 0 & 1/a(x) \end{pmatrix}$ *where $\log|a| \in L^1(\mu)$ and the characteristic function of $E \subset M$ is not cohomologous to zero* $\mod 2$.

(d) $B(x) = \begin{pmatrix} a(x) & 0 \\ 0 & 1/a(x) \end{pmatrix}$ *with $\lambda = \int \log|a|\, d\mu > 0$.*

Moreover, $\|B(x)\|$ and $\|P(f(x))P^{-1}(x)\|$ are bounded above by $\|A(x)\|$. In particular, the Lyapunov exponents of B are well defined and coincide with the Lyapunov exponents of A. The Lyapunov exponents vanish in cases (a)–(c), and they are $\pm\lambda$ in case (d).

In the remainder of the present section we prove Theorem 6.11.

6.3.1 Non-atomic measures

Let $B : \mathbb{R}^2 \to \mathbb{R}^2$ be a non-zero linear map. When B is invertible, we denote by $B_*\theta$ the push-forward of a probability measure θ under the projective action $[v] \mapsto [B(v)]$. When B is non-invertible (of rank 1, necessarily) the projective action is defined at all points except $[\ker B] \in \mathbb{PR}^2$, and so $B_*\theta$ is still defined for every non-atomic probability measure θ. In this case, $B_*\theta$ is the Dirac mass at the image $[B(\mathbb{R}^2)] \in \mathbb{PR}^2$.

Lemma 6.13 *The map $(B,\theta) \mapsto B_*\theta$ is continuous, for B in the space of non-zero linear maps of $\mathbb{R}^2$, with the coefficients topology, and θ in the space of non-atomic probability measures on $\mathbb{PR}^2$, with the weak* topology.*

Proof Let $(B_n)_n$ converge to B and $(\theta_n)_n$ converge to θ. Consider any continuous function $\varphi : \mathbb{PR}^2 \to \mathbb{R}$. Then $|\int(\varphi \circ B_n)\, d\theta_n - \int(\varphi \circ B)\, d\theta|$ is bounded by

$$\left|\int(\varphi \circ B_n)\, d\theta_n - \int(\varphi \circ B)\, d\theta_n\right| + \left|\int(\varphi \circ B)\, d\theta_n - \int(\varphi \circ B)\, d\theta\right|. \quad (6.12)$$

Suppose that B is invertible. Then the projective action of B_n converges to the projective action of B, uniformly on $\mathbb{PR}^2$. The first difference in (6.12) is bounded by $\sup|\varphi \circ B_n - \varphi \circ B|$, which goes to zero when $n \to \infty$. The second also goes to zero, because $\varphi \circ B$ is continuous and θ_n converges to θ. Since φ is arbitrary, this proves that $(B_n)_*\theta_n$ converges to $B_*\theta$ in the weak* topology.

Now suppose B is non-invertible, of rank 1. Given $\varepsilon > 0$, fix a closed neighborhood U of $[\ker B] \in \mathbb{PR}^2$ such that $\theta(U) < \varepsilon$. Then $\theta_n(U) < \varepsilon$ for every large n. The difference $|\int(\varphi \circ B_n)\,d\theta_n - \int(\varphi \circ B)\,d\theta|$ is bounded by

$$\left|\int_{U^c}(\varphi \circ B_n)\,d\theta_n - \int_{U^c}(\varphi \circ B)\,d\theta\right| + \left|\int_{U}(\varphi \circ B_n)\,d\theta_n - \int_{U}(\varphi \circ B)\,d\theta\right|. \quad (6.13)$$

The projective action of B_n converges to the projective action of B uniformly on U^c. So the same argument as before shows that the first difference in (6.13) is less than ε if n is large enough. The second is bounded by $2\varepsilon \sup|\varphi|$. Since ε is arbitrary, this proves that $(B_n)_*\theta_n$ converges to $B_*\theta$ also in this case. □

We use $B^t : \mathbb{R}^d \to \mathbb{R}^d$ to denote the *adjoint* of a linear map $B : \mathbb{R}^d \to \mathbb{R}^d$:

$$B^t(u) \cdot v = u \cdot B(v) \quad \text{for every } u,\, v \in \mathbb{R}^d.$$

Lemma 6.14 *Let θ be a non-atomic probability measure on $\mathbb{PR}^2$ and $(B_n)_n$ be a sequence in* SL(2). *The following conditions are equivalent:*

(1) $\|B_n\| \to \infty$ *as* $n \to \infty$;
(2) *every limit point of $(B_n)_*\theta$ in the weak* topology is a Dirac mass $\delta_{[v]}$.*

If $(B_n)_*\theta \to \delta_{[v]}$ then $\|B_n^t(u)\|/\|B_n\| \to |u \cdot v|/\|v\|$ for every $u \in \mathbb{R}^2$.

Proof Suppose that $\|B_n\|$ does not converge to ∞. Then $(B_n)_n$ admits some subsequence converging to an invertible linear map $B : \mathbb{R}^2 \to \mathbb{R}^2$. Then, by Lemma 6.13, the corresponding push-forwards of θ converge to $B_*\theta$, which is non-atomic. This shows that (2) implies (1). Now let us prove that (1) implies (2) and the last conclusion in the statement. Suppose that $\|B_n\| \to \infty$. Up to restricting to a subsequence, we may suppose that $L_n = B_n/\|B_n\|$ converges to some $L : \mathbb{R}^2 \to \mathbb{R}^2$. Note that $\|L\| = 1$ and $|\det L| = \lim_n 1/\|B_n\|^2 = 0$, and so L has rank 1. Let $V = L(\mathbb{R}^2)$. By Lemma 6.13, the image $(B_n)_*\theta = (L_n)_*\theta$ converges to δ_V. Now fix unit vectors $v \in V$ and $w \in V^\perp$. Any vector $u \in \mathbb{R}^2$ may be written as $u = (u \cdot v)v + (u \cdot w)w$. Then

$$\frac{\|B_n^t u\|}{\|B_n\|} = \|L_n^t(u)\| \to \|L^t(u)\| = |u \cdot v|$$

because $L^t(v) = \pm v$ and $L^t(w) = 0$. □

Corollary 6.15 *The* stabilizer $H(\theta) = \{B \in \mathrm{SL}(2) : B_*\theta = \theta\}$ *of any non-atomic probability measure* θ *on* $\mathbb{PR}^2$ *is a compact subgroup of* $\mathrm{SL}(2)$.

Proof The fact that $H(\theta)$ is a subgroup of $\mathrm{SL}(2)$ is clear from the definition and the fact that it is closed follows from Lemma 6.13. So, we only have to prove that the norm of the elements of H_θ is bounded. That is a simple consequence of Lemma 6.14: if there exists a sequence $A_n \in H(\theta)$, $n \geq 1$ with $\|A_n\| \to \infty$ then the sequence $(A_n)_*\eta$ accumulates on some Dirac mass; that is not possible because this sequence is constant equal to η, which is non-atomic. This contradiction proves that $H(\theta)$ is indeed bounded in norm. □

6.3.2 Convergence to a Dirac mass

The *adjoint* F^t of a linear cocycle $F(x,v) = (f(x), A(x)v)$ is the linear cocycle defined over the inverse $f^{-1} : M \to M$ by $F^t(x,v) = (f^{-1}(x), A^t(x)v)$, where $A^t(x) = A(f^{-1}(x))^t$. The nth iterate is given by $(x,v) \mapsto (f^{-n}(x), A^{tn}(x))$, where

$$\begin{aligned} A^{tn}(x) &= A^t(f^{-n+1}(x)) \cdots A^t(f^{-1}(x)) A^t(x) \\ &= A(f^{-n}(x))^t \cdots A(f^{-2}(x))^t A(f^{-1}(x))^t. \end{aligned}$$

The adjoint F^t is not locally constant, because $A^t(x)$ depends on the first (not the zeroth) coordinate of x. However, F^t is conjugate to the locally constant cocycle $F'(x,v) = (f^{-1}(x), A(x)^t v)$ by the transformation $(x,v) \mapsto (f^{-1}(x), v)$. We denote by $\mathbb{P}F^t$ and $\mathbb{P}F'$ the projective cocycles associated with F^t and F', respectively. The notions of stationary measures for $\mathbb{P}F^t$ and $\mathbb{P}F'$ coincide.

Let η^t be any $\mathbb{P}F^t$-stationary probability measure on $\mathbb{PR}^2$. We are going to analyze the iterates of η^t under the matrices

$$A^n(x)^t = A(x)^t \cdots A(f^{n-1}(x))^t = A^{tn}(f^n(x)).$$

Lemma 6.16 *For* μ*-almost every* $x \in M$*, there exists a probability measure* m_x *on* $\mathbb{PR}^2$ *such that* $A^n(x)^t_*\eta^t \to m_x$ *in the weak* topology.*

Proof This is analogous to Lemma 5.22; we just describe the main steps. Given any continuous function $\varphi : \mathbb{PR}^2 \to \mathbb{R}$ and $n \geq 0$, define

$$\Psi_{\varphi,n}(x) = \int \varphi \circ A^n(x)^t \, d\eta^t.$$

Let $\mathscr{F}_0 = \{\emptyset, M\}$ and, for each $n \geq 1$, let $\mathscr{F}_n$ be the σ-algebra generated by the cylinders $[0 : \Delta_0, \ldots, \Delta_{n-1}]$ with $\Delta_i \subset X$ a measurable set for $i = 0, \ldots, n-1$. Then $\mathscr{F}_n \subset \mathscr{F}_{n+1}$ for every $n \geq 0$. Since $A^n(x)^t$ depends only on $x_0, \ldots, x_{n-1}$, so does the value of $\Psi_{\varphi,n}(x)$. Hence, every $\Psi_{\varphi,n}$ is $\mathscr{F}_n$-measurable. Moreover, for any cylinder $C \in \mathscr{F}_n$,

$$\int_C \Psi_{\varphi,n+1}(x)\,d\mu(x) = \int_C \varphi \circ A^{n+1}(x)^t\,d\eta^t\,d\mu(x)$$
$$= \int_C \varphi \circ A^n(x)^t\,d\eta^t\,d\mu(x)$$

because η^t is $\mathbb{P}F^t$-stationary. This proves that $(\Psi_{\varphi,n},\mathscr{F}_n)$ is a martingale. It follows that there exists a measurable function $\Psi_\varphi : M \to \mathbb{R}$ such that

$$\Psi_\varphi(x) = \lim \Psi_{\varphi,n}(x)$$

for every x in some full μ-measure set $M_0 \subset M$. Using the fact that the space $C^0(\mathbb{PR}^2)$ is separable, one can choose the full measure subset M_0 to be the same for every continuous function φ. By the Riesz representation theorem, each linear map $\varphi \mapsto \Psi_\varphi(x)$, $x \in M_0$ defines a probability measure m_x on $\mathbb{PR}^2$. By construction, m_x is the weak* limit of $A^n(x)^t_*\eta^t$. □

Lemma 6.17 *Let m_x be as in Lemma 6.16. Then $A^n(x)^t_*B^t_*\eta^t \to m_x$ for μ-almost every x and ν-almost every $B \in \mathrm{SL}(2)$.*

Proof Given any continuous function $\varphi : \mathbb{PR}^2 \to \mathbb{R}$ and any $n \geq 0$, define $\Psi_{\varphi,n} : M \to \mathbb{R}$ as in the proof of Lemma 6.16:

$$\Psi_{\varphi,n}(x) = \int \varphi \circ A^n(x)^t\,d\eta^t.$$

We claim that $\int \Psi_{\varphi,n}\Psi_{\varphi,n+1}\,d\mu = \int \Psi^2_{\varphi,n}\,d\mu$ for every $n \geq 0$. Indeed, the expression on the left-hand side may be written as

$$\int \Psi_{\varphi,n}(x)\Big[\int \varphi \circ A^n(x)^t \circ A(f^n(x))^t\,d\eta^t\Big]\,d\mu(x).$$

Since $\Psi_{\varphi,n}(x)$ and $A^n(x)^t$ depend only on $x_0,\ldots,x_{n-1}$, and $A(f^n(x))^t$ depends only on x_n, this is the same as

$$\int \Psi_{\varphi,n}(x)\Big[\int \varphi \circ A^n(x)^t \circ A(f^n(x))^t\,d\eta^t\,dp(x_n)\Big]\,dp^n(x_0,\ldots,x_{n-1}).$$

Using the assumption that η^t is $\mathbb{P}F^t$-stationary, this expression becomes

$$\int \Psi_{\varphi,n}(x)\Big[\int \varphi \circ A^n(x)^t\,d\eta^t\Big]\,dp^n(x_0,\ldots,x_{n-1}) = \int \Psi^2_{\varphi,n}d\mu.$$

That proves our claim. Now it follows that

$$\int \left[\Psi_{\varphi,n+1} - \Psi_{\varphi,n}\right]^2 d\mu = \int \left[\Psi^2_{\varphi,n+1} - 2\int \Psi_{\varphi,n}\Psi_{\varphi,n+1} + \int \Psi^2_{\varphi,n}\right] d\mu$$
$$= \int \Psi^2_{\varphi,n+1}\,d\mu - \int \Psi^2_{\varphi,n}\,d\mu.$$

Therefore, by telescopic cancellation,

$$\sum_{n=0}^{l-1} \int \left[\Psi_{\varphi,n+1} - \Psi_{\varphi,n}\right]^2 d\mu = \int \Psi_{\varphi,l}^2 \, d\mu - \int \Psi_{\varphi,0}^2 \, d\mu \le \sup|\varphi|^2 \tag{6.14}$$

for every $l \ge 1$. The integral on the left-hand side may be rewritten as

$$\int \Big[\int \varphi \circ A^n(x)^t A(f^n(x))^t \, d\eta^t - \int \varphi \circ A^n(x)^t \, d\eta^t\Big]^2 dp^n(x_0,\dots,x_{n-1}) \, dp(x_n);$$

that is,

$$\begin{aligned}\int \Big[\int \varphi \circ A^n(x)^t B^t \, d\eta^t \; - \int \varphi \circ A^n(x)^t \, d\eta^t\Big]^2 dp^n(x_0,\dots,x_{n-1}) \, d\nu(B) \\ = \int \Big[\int \varphi \circ A^n(x)^t B^t \, d\eta^t \; - \int \varphi \circ A^n(x)^t \, d\eta^t\Big]^2 d\mu(x) \, d\nu(B).\end{aligned}$$

Thus, (6.14) means that

$$\sum_{n=0}^{\infty} \int \Big[\int \varphi \circ A^n(x)^t B^t \, d\eta^t \; - \int \varphi \circ A^n(x)^t \, d\eta^t\Big]^2 d\mu(x) \, d\nu(B) < \infty.$$

By Borel–Cantelli, this implies that

$$\lim_n \Big[\int \varphi \circ A^n(x)^t B^t \, d\eta^t \; - \int \varphi \circ A^n(x)^t \, d\eta^t\Big] = 0 \tag{6.15}$$

for $(\mu \times \nu)$-almost every (x,B). Taking the intersection of those full measure sets over all functions φ in some countable dense subset of $C^0(\mathbb{P}\mathbb{R}^2)$, we find a full $(\mu \times \nu)$-measure set of pairs (x,B) such that (6.15) holds; that is,

$$\lim_n \int \varphi \circ A^n(x)^t B^t \, d\eta^t = \int \varphi \, dm_x,$$

for all continuous functions $\varphi : \mathbb{P}\mathbb{R}^2 \to \mathbb{R}$. This is the claim of the lemma. □

Lemma 6.18 *For μ-almost any x there is $V(x) \in \mathbb{P}\mathbb{R}^2$ such that $m_x = \delta_{V(x)}$.*

Proof By Lemma 6.9, the $\mathbb{P}F^t$-stationary measure η^t is non-atomic. Hence, by Corollary 6.15, the stabilizer $H(\eta^t)$ is a compact subgroup of SL(2). By Lemma 6.17, for μ-almost every $x \in M$,

$$\lim_n A^n(x)^t_* B^t_* \eta^t = \lim_n A^n(x)^t_* \eta^t = m_x \quad \text{for } \nu\text{-almost every } B.$$

Fix x and let P be any accumulation point of the sequence $A^n(x)^t / \|A^n(x)^t\|$. Then $\|P\| = 1$ and, by Lemma 6.13, the previous relation implies

$$P_* B^t_* \eta^t = P_* \eta^t = m_x \quad \text{for } \nu\text{-almost every } B^t.$$

If P was invertible then $B^t_* \eta^t = \eta^t$ for ν-almost every B, contradicting the fact that the stabilizer of η^t is compact. Thus, P has rank 1 and so $m_x = P_* \eta^t$ is the Dirac measure $\delta_{V(x)}$ with $V(x) = P(\mathbb{R}^2)$. □

6.3.3 Proof of Theorem 6.11

Let η be any $\mathbb{P}F$-stationary measure. By Theorem 6.8, the largest Lyapunov exponent λ_+ is given by

$$\lambda_+ = \int \Phi \, d\eta \, d\mu. \tag{6.16}$$

Our goal is to show that the integral is positive. Let $V(x)$ be as in Lemma 6.18.

Corollary 6.19 *$\|A^n(x)u\| \to \infty$ for every $u \notin V(x)^\perp$ and μ-almost every $x \in M$.*

Proof We have seen in Lemma 6.18 that $A_n(x)^t_* \eta^t$ converges to a Dirac mass, at μ-almost every point. Write $V(x) = [v]$. Then Lemma 6.14 gives that

$$\|A^n(x)^t\| \to \infty \quad \text{and} \quad \frac{\|A^n(x)u\|}{\|A^n(x)^t\|} \to \frac{|u \cdot v|}{\|v\|}$$

for every u. In particular, $\|A^n(x)u\| \to \infty$ for every $u \notin V(x)^\perp$. □

Since η is non-atomic, by Lemma 6.9, the set $\{(x, V(x)) : x \in M\}$ has zero $\mu \times \eta$-measure. So, Corollary 6.19 implies that

$$\lim_n \sum_{j=0}^{n} \Phi(F^j(x,[v])) = \lim_n \log \frac{\|A^n(x)v\|}{\|v\|} = \infty \tag{6.17}$$

for $\mu \times \eta$-almost every $(x,[v])$. We need the following abstract result:

Lemma 6.20 *Let $(X, \mathscr{X}, m)$ be a probability space and $T : X \to X$ a measure-preserving map. If $\varphi \in L^1(X,m)$ is such that*

$$\lim_n \sum_{j=0}^{n-1} \varphi(T^j(x)) = +\infty \quad \textit{for } m\textit{-almost every } x \in X,$$

then $\int \varphi \, dm > 0$.

Proof Let $S_n = \sum_{j=0}^{n-1} \varphi \circ T^j$ and, for each fixed $\varepsilon > 0$, define

$$A_\varepsilon = \{x \in X : S_n(x) > \varepsilon \text{ for all } n \geq 1\} \text{ and } B_\varepsilon = \bigcup_{k \geq 0} T^{-k}(A_\varepsilon).$$

We claim that if $S_n(x) \to \infty$ then x belongs to B_ε for some $\varepsilon > 0$. Indeed, suppose otherwise. Then $x \notin B_{1/l^2}$ for any $l \geq 1$. This means that $T^k(x) \notin A_{1/l^2}$ for any $k \geq 0$ and $l \geq 1$. Equivalently, for any $k \geq 0$ and $l \geq 1$ there exists $n \geq 1$ such that $S_n(f^k(x)) \leq 1/l^2$. Then we may construct a sequence $(n_l)_l$ with $n_0 = 0$ and

$$S_{n_l}(f^{n_+ \cdots + n_{l-1}}(x)) \leq 1/l^2 \quad \text{for every } l \geq 1.$$

By construction, $S_{n_1+\cdots+n_l}(x) \le 1 + 1/4 + \cdots + 1/l^2$ does not converge to ∞. This proves our claim. It follows from the hypothesis of the lemma that the union of all B_ε, $\varepsilon > 0$ has full measure.

Now let ψ be the Birkhoff average of φ and let τ_ε be the sojourn time in A_ε; that is, the Birkhoff average of the characteristic function $\mathscr{X}_{A_\varepsilon}$. The hypothesis of the lemma implies that $\psi(x) \ge 0$ for almost every x, and so $\int \varphi\, dm = \int \psi\, dm \ge 0$. Suppose, for contradiction, that the integrals vanish. Then $\psi = 0$ almost everywhere. Given $x \in B_\varepsilon$, let $k \ge 0$ be the smallest integer such that $T^k(x) \in A_\varepsilon$. Then

$$\sum_{j=0}^{n-1} \varphi(T^j x) \ge \sum_{j=0}^{k-1} f(T^j x) + \sum_{j=k}^{n-1} \varepsilon \mathscr{X}_{A_\varepsilon}(T^j x), \text{ for all } n \ge 1.$$

Dividing by n and sending $n \to \infty$, we obtain $0 = \psi(x) \ge \varepsilon \tau_\varepsilon(x)$ for every $x \in B_\varepsilon$. It is clear that $\tau_\varepsilon(x)$ for every $x \notin B_\varepsilon$. So, we conclude that $m(A_\varepsilon) = \int \tau_\varepsilon\, dm = 0$. Then $m(B_\varepsilon)$ also vanishes, for every $\varepsilon > 0$. This contradicts the claim in the previous paragraph. This contradiction proves the lemma. □

In view of (6.16) and (6.17), Lemma 6.20 gives that $\lambda_+ = \int \Phi\, d\eta\, d\mu$ is strictly positive. This finishes the proof of Theorem 6.11.

6.4 Notes

Our presentation in Section 6.1 is a variation of the original proof of Theorem 6.1, which was due to Ledrappier [80, § I.5].

Theorem 6.8 is taken from Furstenberg [54]. Corollary 6.10 is contained in results of Furstenberg, and Kifer [57] and of Hennion [61]. Related results, for some reducible cocycles, were obtained by Kifer and Slud [70, 73]. We will return to the topic of continuity of Lyapunov exponents in Chapter 10.

Theorem 6.11 is also due to Furstenberg [54]. Several extensions have been obtained, by Virtser [116], Guivarc'h [59], Royer [103], Ledrappier and Royer [82] and others. This includes the proof that the statement remains true when the base dynamics (f, μ) is just a Markov shift.

More recently, Bonatti, Gomez-Mont and Viana [37] extended the criterion to Hölder-continuous cocycles with invariant holonomies over hyperbolic homeomorphisms. The invariant measure only needs to have local product structure. Currently, the strongest statements are due to Viana [113], for Hölder-continuous cocycles over (non-uniformly) hyperbolic systems, and Avila, Santamaria and Viana [13], when the base dynamics is partially hyperbolic.

Two different generalizations of Furstenberg's theorem are at the heart of the next two chapters: the extension of the Furstenberg criterion to arbitrary dimension, which we discuss in Chapter 7, and the criterion for simplicity of the whole Lyapunov spectrum, to be presented in Chapter 8.

6.5 Exercises

Exercise 6.1 Let $Y \subset X$ be either open or closed. Suppose that $(\mu_n)_n$ converges to μ in the weak* topology on X and let $(\mu_{Y,n})_n$ and μ_Y be their normalized restrictions to Y. Conclude that $(\mu_{Y,n})_n$ converges to μ_Y in the weak* topology on Y.

Exercise 6.2 Use Exercise 5.22 to check that for each $j = 1, \dots, k$, the probability measure m in the second part of Theorem 6.1 may be chosen to be ergodic.

Exercise 6.3 Use Exercise 5.23 to check that Proposition 6.5 remains true if one restricts to ergodic s-states and ergodic u-states.

Exercise 6.4 Show that Proposition 6.7 and the expressions (a) and (b) for the Lyapunov exponents remain true if one restricts to ergodic stationary measures.

Exercise 6.5 Suppose that $X = \mathrm{SL}(2)$ and the probability measure p has finite support; that is, it is of the form $p = p_1\delta_{A_1} + \cdots + p_m\delta_{A_m}$. Moreover, let $A : M \to \mathrm{SL}(2)$ be the projection to the zeroth coordinate. Show that the associated linear cocycle is strongly irreducible if and only if it is twisting.

Exercise 6.6 Prove that the following conditions are equivalent.

(1) The support of ν is not contained in any compact subgroup of $\mathrm{SL}(2)$ (hypothesis (a) in Theorem 6.11).
(2) The support of ν is not contained in a set of the form $G\mathscr{R}G^{-1}$ where $G \in \mathrm{SL}(2)$ and $\mathscr{R} \subset \mathrm{SL}(2)$ is the group of rotations.
(3) The linear cocycle F is pinching: the monoid $\mathscr{B}$ generated by $\operatorname{supp}\nu$ contains matrices with arbitrarily large norm.

Exercise 6.7 Let F be pinching. Prove that the following conditions are equivalent.

(1) There is no finite set $L \subset \mathbb{PR}^2$ such that $B(L) = L$ for ν-almost every $B \in \mathrm{SL}(2)$ (hypothesis (b) in Theorem 6.11).
(2) There is no set $L \subset \mathbb{PR}^2$ with one or two elements such that $B(L) = L$ for ν-almost every $B \in \mathrm{SL}(2)$.

(3) The linear cocycle F is twisting: for any $V_0, V_1, \ldots, V_k \in \mathbb{PR}^2$ there is $B \in \mathscr{B}$ such that $B(V_0) \neq V_i$ for all $i = 1, \ldots, k$.

Exercise 6.8 Calculate the stationary measures for the product of random matrices in the second part of Exercise 1.2. Deduce that this cocycle is a point of continuity for the Lyapunov exponents among locally constant cocycles.

Exercise 6.9 Let F be an SL(2)-cocycle. Use Lemma 6.9 and Corollary 6.15 to show that if F satisfies conditions (a) and (b) in Theorem 6.11 then there exists no probability measure on $\mathbb{PR}^2$ such that $B_*\eta = \eta$ for every $B \in \operatorname{supp} \nu$.

Exercise 6.10 Let F be an SL(2)-cocycle. Prove the converse to Exercise 6.9 and deduce that if $\lambda_\pm = 0$ then there exists some probability measure η on $\mathbb{PR}^2$ such that $B_*\eta = \eta$ for every $B \in \operatorname{supp} \nu$.

Exercise 6.11 Show that the cocycle F' in Section 6.3.2 satisfies the hypotheses of Theorem 6.11 if and only if F does.

7

Invariance principle

The main result in this chapter takes our understanding of the invariant measures of linear and projective cocycles, and their links to Lyapunov exponents, one step further. Most important, it applies precisely when the information provided by the theorem of Oseledets is poor, namely, when all Lyapunov exponents coincide and so the Oseledets decomposition is trivial.

As a motivation for the statement, let us start with a brief discussion of the 2-dimensional case. Let $F : M \times \mathbb{R}^2 \to M \times \mathbb{R}^2$ be an invertible locally constant linear cocycle over a Bernoulli shift $f : (M, \mu) \to (M, \mu)$, and let $\mathbb{P}F : M \times \mathbb{P}\mathbb{R}^2 \to M \times \mathbb{P}\mathbb{R}^2$ be the associated projective cocycle.

Suppose first that the Lyapunov exponents $\lambda_\pm$ of F are distinct. Then, as we have seen in Section 5.5.3, there exists an s-state m^s and a u-state m^u such that every $\mathbb{P}F$-invariant probability measure that projects down to μ is a convex combination of m^s and m^u. Moreover, these are the unique $\mathbb{P}F$-ergodic probability measures, and they determine the Lyapunov exponents:

$$\lambda_+ = \int \Phi \, dm^u \quad \text{and} \quad \lambda_- = \int \Phi \, dm^s. \tag{7.1}$$

When the Lyapunov exponents coincide, we still get that $\lambda_\pm = \int \Phi \, dm$ for every $\mathbb{P}F$-invariant probability measure. Moreover, the *invariance principle* that we are going to prove in this chapter provides a surprisingly precise characterization of these measures: every $\mathbb{P}F$-invariant probability measure that projects down to μ is a product measure $m = \mu \times \eta$, where η is both forward and backward stationary.

The invariance principle holds in arbitrary dimension, as we are going to see in Sections 7.1 and 7.2. In Section 7.3 we deduce the following result, which was originally due to Furstenberg [54]: if $\lambda_- = \lambda_+$ then there exists some probability measure η on projective space that is invariant under $A(x)$ for almost every x. In Section 7.4 we observe that the latter is a very restrictive condition on the cocycle. Thus, one has $\lambda_- < \lambda_+$ for generic locally constant cocycles.

7.1 Statement and proof

Let $\hat{F} : \hat{M} \times \mathbb{R}^d \to \hat{M} \times \mathbb{R}^d$ be an invertible linear cocycle defined over a two-sided Bernoulli shift $\hat{f} : (\hat{M}, \hat{\mu}) \to (\hat{M}, \hat{\mu})$ by a measurable $\hat{A} : \hat{M} \to \mathrm{GL}(d)$ that depends only on the zeroth coordinate. Let $\mathbb{P}\hat{F} : \hat{M} \times \mathbb{PR}^d \to \hat{M} \times \mathbb{PR}^d$ be the associated projective cocycle. Write $\hat{M} = X^{\mathbb{Z}}$, where X and $\hat{M}$ are taken to be separable complete metric spaces and $\hat{\mu} = p^{\mathbb{Z}}$. Assume that $\log^+ \|\hat{A}^{\pm 1}\| \in L^1(\hat{\mu})$ and let λ_- and λ_+ be the extremal Lyapunov exponents of $\hat{F}$.

Theorem 7.1 (Linear invariance principle) *If $\lambda_- = \lambda_+$ then every $\mathbb{P}\hat{F}$-invariant probability measure $\hat{m}$ that projects down to $\hat{\mu}$ is both an s-state and a u-state. Consequently, $\hat{m} = \hat{\mu} \times \eta$ for some probability measure η on $\mathbb{PR}^d$ that is both forward and backward stationary.*

Theorem 7.1 is a consequence of the following theorem of Ledrappier [81]. Let $F : M \times \mathbb{R}^d \to M \times \mathbb{R}^d$ be a linear cocycle defined over a measure-preserving transformation $f : (M, \mu) \to (M, \mu)$ by a measurable function $A : M \to \mathrm{GL}(d)$ such that $\log^+ \|A^{\pm 1}\| \in L^1(\mu)$. Let $\mathbb{P}F : M \times \mathbb{PR}^d \to M \times \mathbb{PR}^d$ be the associated projective cocycle. Note F, $\mathbb{P}F$, and f are not assumed to be invertible. Moreover, f is arbitrary (not necessarily a shift) and A need not be locally constant either. We do assume that M is a separable complete metric space (and so (M, μ) is a Lebesgue space, see [114, Section 8.5.2]).

Theorem 7.2 (Ledrappier) *Assume that $\lambda_-(F, x) = \lambda_+(F, x)$ for μ-almost every $x \in M$. Then*

$$m_{f(x)} = A(x)_* m_x \quad \textit{for } \mu\textit{-almost every } x \in M,$$

for any disintegration $\{m_x : x \in M\}$ of any $\mathbb{P}F$-invariant probability measure m that projects down to μ.

Proof of Theorem 7.1 Let $M = X^{\mathbb{N}}$, $\mu = p^{\mathbb{N}}$ and $f : (M, \mu) \to (M, \mu)$ be the one-sided Bernoulli shift. Define $A : M \to \mathrm{GL}(d)$ by $\hat{A} = A \circ \pi^+$ and then take $F : M \times \mathbb{R}^d \to M \times \mathbb{R}^d$ to be the linear cocycle and $\mathbb{P}F : M \times \mathbb{PR}^d \to M \times \mathbb{PR}^d$ to be the projective cocycle defined by A over f. Then $\lambda_-(F, x) = \lambda_-$ and $\lambda_+(F, x) = \lambda_+$ for μ-almost every $x \in M$. Given any $\mathbb{P}\hat{F}$-invariant probability measure $\hat{m}$, let $m = \Pi^+_* \hat{m}$ be its projection to $M \times \mathbb{PR}^d$. We know from Corollary 5.23 that the disintegrations $\{\hat{m}_{\hat{x}} : \hat{x} \in \hat{M}\}$ and $\{m_x : x \in M\}$ of $\hat{m}$ and m, respectively, are related by

$$\hat{m}_{\hat{x}} = \lim_n \hat{A}^n(\hat{f}^{-n}(\hat{x}))_* m_{\pi^+(\hat{f}^{-n}(\hat{x}))} \quad \text{for } \hat{\mu}\text{-almost every } \hat{x} \in \hat{M}.$$

Now, using Theorem 7.2 and the relations $f \circ \pi^+ = \pi^+ \circ f$ and $\hat{A} = A \circ \pi^+$,

$$\begin{aligned}\hat{A}^n(\hat{f}^{-n}(\hat{x}))_* m_{\pi^+(\hat{f}^{-n}(\hat{x}))} &= A^n(\pi^+(\hat{f}^{-n}(\hat{x})))_* m_{\pi^+(\hat{f}^{-n}(\hat{x}))} \\ &= m_{f^n(\pi^+(\hat{f}^{-n}(\hat{x})))} = m_{\pi^+(x)}\end{aligned}$$

for $\hat{\mu}$-almost every $\hat{x}$. Substituting this in the previous expression we conclude that $\hat{m}_{\hat{x}} = m_{\pi^+(x)}$ for $\hat{\mu}$-almost every $\hat{x}$. In particular, $\hat{m}_{\hat{x}}$ depends only on $\pi^+(x)$, and that proves that $\hat{m}$ is an s-state. An entirely dual argument, using backward iterates instead, proves that $\hat{m}$ is a u-state. □

7.2 Entropy is smaller than exponents

In this section we prove Theorem 7.2. Let m be a probability measure on $M \times \mathbb{PR}^d$ that projects down to μ and let $\{m_x : x \in M\}$ be a disintegration of m. For each x, let $J(x,\cdot)$ be the Radon–Nikodym derivative

$$J(x,v) = \frac{d\big(A(x)_*^{-1} m_{f(x)}\big)}{dm_x}(v).$$

This means that

$$A(x)_*^{-1} m_{f(x)} = J(x,\cdot) m_x + \xi_x \tag{7.2}$$

for some positive measure ξ_x (totally) singular with respect to m_x. The *fibered entropy* of m is defined by

$$h(m) = \int -\log J\, dm = \int_{\{J>0\}} -\log J\, dm + \infty m(\{J=0\}). \tag{7.3}$$

Note that $\int_{\{J>0\}} J\, dm = \int J\, dm \le 1$, because the left-hand side of (7.2) is a probability measure and ξ_x is a positive measure. Then, by convexity,

$$\int_{\{J>0\}} -\log J\, dm \ge -\log \int_{\{J>0\}} J\, dm \ge 0. \tag{7.4}$$

This shows that the integral in (7.3) is well defined and $h(m) \in [0,+\infty]$.

Proposition 7.3 *If $h(m) = 0$ then $A(x)_* m_x = m_{f(x)}$ for μ-almost every x.*

Proof The first inequality in (7.4) is strict unless $\log J$ is constant m-almost everywhere. The second one is strict unless $\int J\, dm = 1$ (equivalently, $\xi_x = 0$ for μ-almost every x). Therefore, $h(m) = 0$ if and only if $m(\{J=0\}) = 0$, and $\int J\,dm = 1$, and $\log J$ is constant on a full m-measure set. In particular, if $h(m) = 0$ then $J = 1$ on a full m-measure set. That is, $A(x)_*^{-1} m_{f(x)} = m_x$ or, equivalently, $A(x)_* m_x = m_{f(x)}$ for μ-almost every $x \in M$. This proves the claim. □

Proposition 7.4 $0 \le h(m) \le d(\lambda_+ - \lambda_-)$.

Theorem 7.2 is a direct consequence of Propositions 7.3 and 7.4. In the remainder of this section we prove Proposition 7.4.

7.2.1 The volume case

As a motivation for the proof, and for the definition of $h(m)$, we begin by discussing the case when $m = \mu \times \lambda$, where λ is the Haar measure on $\mathbb{PR}^d$. This will not be used in the proof of the general case, so the reader may choose to skip this section altogether.

In the case under consideration $m_x = \lambda$ for every $x \in M$, and so the Radon–Nikodym derivative

$$J(x,[v]) = \frac{A(x)_*^{-1}\lambda}{d\lambda}([v]) = |\det D\Phi_x([v])|$$

where $\Phi_x : \mathbb{PR}^d \to \mathbb{PR}^d$, $\Phi_x([v]) = [A(x)v]$ denotes the projective action of $A(x)$.

It follows that $J(x,[v]) = |\det A(x)|\left(\|v\|/\|A(x)v\|\right)^d$ at every point, and so

$$\begin{aligned} h(m) &= d\int \log\frac{\|A(x)v\|}{\|v\|}\,d\mu(x)\,d\lambda([v]) - \int \log|\det A(x)|\,d\mu(x) \\ &= d\int \Phi\,d\mu\,d\lambda - \int \log|\det A(x)|\,d\mu(x), \end{aligned}$$

where Φ is the function in (6.1). Let $m = \int m_\alpha\,d\alpha$ be the ergodic decomposition of $m = \mu \times \lambda$. By Theorem 6.1, for each α there exists $j = j(\alpha)$ such that $\int \Phi\,dm_\alpha = \lambda_j$. Thus,

$$\int \Phi\,d\mu\,d\lambda = \sum_{j=1}^{k} c_j\lambda_j, \qquad \sum_{j=1}^{k} c_j = 1,$$

where c_j is the total weight of ergodic components with $j(\alpha) = j$. By (4.40),

$$\int \log|\det A(x)|\,d\mu(x) = \sum_{j=1}^{k} \lambda_j \dim E^j.$$

Therefore,

$$h(m) = \sum_{j=1}^{k} (dc_j - \dim E^j)\lambda_j \le d(\lambda_+ - \lambda_-). \tag{7.5}$$

The last inequality is, obviously, not sharp.

7.2.2 Proof of Proposition 7.4.

Consider any $\Delta > \lambda_+ - \lambda_- \ge 0$. We want to prove that $h(m) \le d\Delta$ for any F-invariant probability measure m that projects down to μ. For the time being, let us assume that m is ergodic for F; the general case will be deduced at the end.

Given $\varepsilon > 0$, define $h_\varepsilon(m) = \int -\log J_\varepsilon \, dm$ where $J_\varepsilon = J + \varepsilon$. By the monotone convergence theorem, $h_\varepsilon(m)$ increases to $h(m)$ as ε decreases to 0. Suppose, for contradiction, that $h(m) > d\Delta$. Then, for any $\varepsilon > 0$ small enough,

$$h_\varepsilon(m) > d\Delta + 4\varepsilon. \tag{7.6}$$

Lemma 7.5 *Each fiber $\{x\} \times \mathbb{PR}^d$ admits partitions $\mathscr{P}_n(x)$, defined for every large n, such that*

(1) $\#\mathscr{P}_n(x) \le e^{n(d\Delta+2\varepsilon)}$,
(2) $\operatorname{diam} \mathscr{P}_n(x) \le e^{-n(\Delta+2\varepsilon)}$,
(3) $m_x(\partial \mathscr{P}_n(x,v)) = 0$ *for every* $v \in \mathbb{PR}^d$.

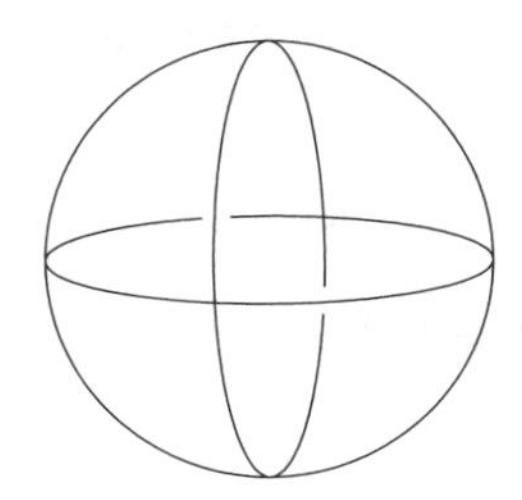
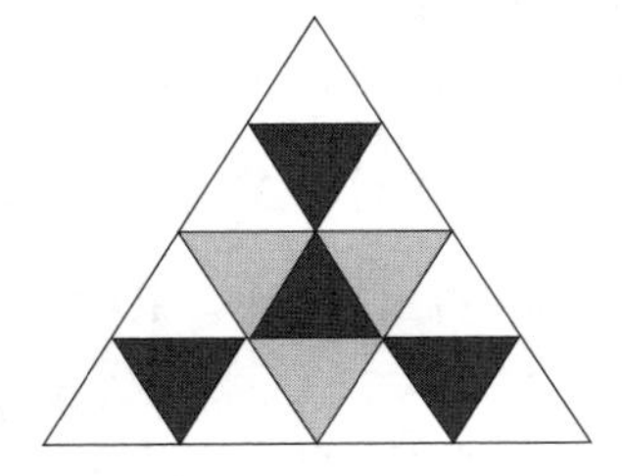

Figure 7.1 A refining sequence of partitions of $\mathbb{PR}^d$

Proof We start from a regular triangulation of the sphere S^{d-1} into 2^d simplices, as pictured in Figure 7.1 for the case $d = 3$. We take the sphere to carry the standard metric with curvature $\equiv 1$. By identifying antipodes, one obtains a triangulation $\mathscr{T}_1$ of the projective space $\mathbb{PR}^d$ into 2^{d-1} simplices. Each simplex can be split into 2^{d-1} regular simplices, each of which is the image of the original simplex by a $(1/2)$-contraction, as illustrated in Figure 7.1. By repeating this procedure successively, we obtain an increasing sequence $\mathscr{T}_k$ of triangulations of $\mathbb{PR}^d$ satisfying

(a) $\#\mathscr{T}_k = 2^{k(d-1)}$ for every $k \geq 1$
(b) $C^{-1} \leq 2^k \operatorname{diam} \mathscr{T}_k(v) \leq C$ for every $k \geq 1$ and $v \in \mathbb{PR}^d$,

where the constant $C > 1$ is independent of k and v.

By construction, the boundaries of the atoms of every $\mathscr{T}_k$ are contained in finitely many projective hyperplanes. Thus, for each $x \in M$ and $k \geq 1$, we may find an orthogonal transformation $B_{x,k} : \mathbb{R}^d \to \mathbb{R}^d$ such that the partition $\mathscr{T}_{x,k} = B_{x,k}(\mathscr{T}_k)$ satisfies

(c) $m_x(\mathscr{T}_{x,k}(v)) = 0$ for every $v \in \mathbb{PR}^d$.

Moreover, properties (a) and (b) are not affected when one replaces $\mathscr{T}_k$ by $\mathscr{T}_{x,k}$, because the projective action of $B_{x,k}$ is an isometry. Assuming ε is small enough relative to Δ, for every large $n \geq 1$ there exists some integer $k \geq 1$ such that

$$n\frac{d\Delta + 2\varepsilon}{(d-1)\log 2} \geq k \geq n\frac{\Delta + 2\varepsilon}{\log 2} + \frac{\log C}{\log 2}.$$

Then the partition $\mathscr{P}_n(x) = \mathscr{T}_{x,k}$ satisfies the conditions in the conclusion of the lemma. □

Let $\mathscr{P}_n(x,v)$ denote the atom of the partition $\mathscr{P}_n(x)$ that contains the point v. For each $0 \leq k \leq n$ let $\mathscr{P}_{n,k}(x)$ be the partition of $\{x\} \times \mathbb{PR}^d$ given by the pull-back of $\mathscr{P}_n(f^k(x))$ under $A^k(x)$:

$$\mathscr{P}_{n,k}(x,v) = A^k(x)^{-1}\big(\mathscr{P}_n(F^k(x,v))\big) \quad \text{for each } v \in \mathbb{PR}^d.$$

Observe that $\mathscr{P}_{n,0}(x,v) = \mathscr{P}_n(x,v)$. Then define, for $0 \leq k < n$,

$$J_{n,k}(x,v) = \frac{m_{f(x)}\big(\mathscr{P}_{n,k}(F(x,v))\big)}{m_x\big(\mathscr{P}_{n,k+1}(x,v)\big)}$$

$$J_n(x,v) = \frac{m_{f^n(x)}\big(\mathscr{P}_n(F^n(x,v))\big)}{m_x\big(\mathscr{P}_{n,n}(x,v)\big)} = \prod_{k=0}^{n-1} J_{n,k}(x,v)$$

and, for each $\varepsilon > 0$,

$$J_{n,k,\varepsilon}(x,v) = J_{n,k}(x,v) + \varepsilon \quad \text{and} \quad J_{n,\varepsilon}(x,v) = \prod_{k=0}^{n-1} J_{n,k}(x,v).$$

Lemma 7.6 $\sup_{0 \leq k \leq n} \| \log J_{n,k,\varepsilon} - \log J_\varepsilon \|_{L^1(m)}$ *converges to zero as* $n \to \infty$.

Let us assume this fact for a while and use it to complete the proof of Proposition 7.4. Since we assume m to be ergodic,

$$\begin{aligned}\lim_n \frac{1}{n} \log J_{n,\varepsilon} &= \lim_n \frac{1}{n} \sum_{k=0}^{n-1} \log J_{n,k,\varepsilon} \circ F^j = \lim_n \frac{1}{n} \sum_{k=0}^{n-1} \log J_\varepsilon \circ F^j \\ &= \int \log J_\varepsilon \, dm = h_\varepsilon(m)\end{aligned}$$

at m-almost every point (the second equality uses Lemma 7.6). Consequently,

$$\begin{aligned}\limsup_n \frac{1}{n} \log m_{f^n(x)}\big(\mathscr{P}_n(F^n(x,v))\big) &\leq \limsup_n \frac{1}{n} \log J_n(x,v) \\ &\leq \lim_n \frac{1}{n} \log J_{n,\varepsilon}(x,v) = -h_\varepsilon(m)\end{aligned}$$

for m-almost every (x,v). In particular, for every large $n \geq 1$ there exists $E_n \subset M \times \mathbb{PR}^d$ such that $m(E_n) > 1/2$ and

$$m_{f^n(x)}\big(\mathscr{P}_n(F^n(x,v))\big) \leq e^{n(-h_\varepsilon(m)+\varepsilon)} \quad \text{for all } (x,v) \in E_n.$$

Since every $\mathscr{P}_n(y)$ has at most $e^{n(d\Delta+2\varepsilon)}$ atoms (Lemma 7.5), and recalling (7.6), it follows that the $m_{f^n(x)}$-measure of the intersection of $F^n(E_n)$ with the fiber of $f^n(x)$ does not exceed

$$e^{n(d\Delta+2\varepsilon)} e^{n(-h_\varepsilon(m)+\varepsilon)} \leq e^{-n\varepsilon}.$$

Since x is arbitrary, we conclude that $m(F^n(E_n)) \leq e^{-n\varepsilon}$, contradicting the fact that $m(E_n) > 1/2$ for every n. This contradiction reduces the proof of Proposition 7.4 to proving Lemma 7.6.

Lemma 7.7 $\sup_{0 \leq k \leq n} \sup\{\operatorname{diam} \mathscr{P}_{n,k}(x,v) : v \in \mathbb{PR}^d\}$ *converges to zero when* $n \to \infty$, *for* μ*-almost every* $x \in M$.

Proof The derivative of the action of $A^k(x)^{-1}$ in projective space is bounded by $\|A^k(x)\| \|A^k(x)^{-1}\|$ (Exercise 7.1). Hence, using Lemma 7.5,

$$\begin{aligned}\operatorname{diam} \mathscr{P}_{n,k}(x,v) &\leq \|A^k(x)\| \|A^k(x)^{-1}\| \operatorname{diam} \mathscr{P}_n(x,v) \\ &\leq \|A^k(x)\| \|A^k(x)^{-1}\| e^{-n(\Delta+2\varepsilon)}.\end{aligned}$$

For μ-almost every $x \in M$,

$$\lim_n \frac{1}{n} \log \|A^k(x)\| \|A^k(x)^{-1}\| = \lambda_+ - \lambda_-.$$

Hence, there exists $k_0(x) \geq 1$ such that

$$\operatorname{diam} \mathscr{P}_{n,k}(x,v) \leq e^{k\Delta} e^{-n(\Delta+2\varepsilon)} \leq e^{-2n\varepsilon} \quad \text{for every } k_0(x) < k \leq n.$$

Fix $k_0(x)$ and let $K_0(x) = \max\{\|A^k(x)\|\|A^k(x)^{-1}\| : 0 \le k \le k_0(x)\}$. Then

$$\operatorname{diam} \mathscr{P}_{n,k}(x,v) \le K_0(x) e^{-n(\Delta+2\varepsilon)} \le e^{-2n\varepsilon}$$

for every n sufficiently large. This proves that, for μ-almost every $x \in M$, there exists $n_0(x) \ge 1$ such that $\sup_{0 \le k \le n} \sup_v \operatorname{diam} \mathscr{P}_{n,k}(x,v) \le e^{-n2\varepsilon}$ for every $n \ge n_0(x)$. That implies the conclusion of the lemma. □

Lemma 7.8 *Let X be a complete metric space and η_0, η_1 be Borel probability measures on X such that $\eta_1 \ge c\eta_0$. Let $\rho : X \to [0, +\infty]$ be the Radon-Nikodym derivative $\rho = d\eta_1/d\eta_0$ and, given any countable partition $\mathscr{P}$ of X, define $\rho_{\mathscr{P}}(x) = \eta_1(\mathscr{P}(x))/\eta_0(\mathscr{P}(x))$ for $x \in X$. Then:*

(1) $\int \log \rho \, d\eta_0 \le \int \log \rho_{\mathscr{P}} \, d\eta_0 \le 0$;
(2) *given $\varepsilon > 0$ there is $\delta > 0$ such that $\|\log \rho_{\mathscr{P}} - \log \rho\|_{L^1(\eta_0)} \le \varepsilon$ for any countable partition $\mathscr{P}$ with $\eta_0(\partial \mathscr{P}) = 0$ and* $\operatorname{diam} \mathscr{P} < \delta$.

Proof By convexity, $\int \log \rho_{\mathscr{P}} \, d\eta_0 \le \log \int \rho_{\mathscr{P}} \, d\eta_0 = 0$. Similarly,

$$\int_{P(x)} \log \rho \, d\eta_0 \le \eta_0(P(x)) \log \rho_{\mathscr{P}}(x)$$

for every atom $P(x)$, and so $\int \log \rho \, d\eta_0 \le \int \log \rho_{\mathscr{P}} \, d\eta_0$. This proves part (1) of the lemma. Next, note that the functions $\rho_{\mathscr{P}}$ satisfy a uniform integrability condition: for all $Y \subset X$ with $\eta_0(Y) < 1/e$,

$$\int_Y |\log \rho_{\mathscr{P}}| \, d\eta_0 \le -\eta_0(Y)(\log \eta_0(Y) + \log c). \tag{7.7}$$

Indeed, the assumption implies $-\log \rho_{\mathscr{P}} \le -\log c$ and so the claim is trivial if $\log \rho_{\mathscr{P}}$ happens to be negative on Y. When $\log \rho_{\mathscr{P}} \ge 0$ on the set Y, the claim follows from convexity:

$$\int_Y \log \rho_{\mathscr{P}} d\frac{\eta_0}{\eta_0(Y)} \le \log \int_Y \rho_{\mathscr{P}} d\frac{\eta_0}{\eta_0(Y)} \le \log \frac{\eta_1(\mathscr{P}(Y))}{\eta_0(Y)} \le \log \frac{1}{\eta_0(Y)},$$

where $\mathscr{P}(Y)$ denotes the union of all atoms of $\mathscr{P}$ that intersect Y. The general case of (7.7) is handled by splitting Y into two subsets where $\log \rho_{\mathscr{P}}$ has constant sign. Next, we claim that if $\mathscr{R}$ refines $\mathscr{Q}$ then

$$\|\log \rho_{\mathscr{R}} - \log \rho_{\mathscr{Q}}\|_{L^1(\eta_0)} \le \|\log \rho - \log \rho_{\mathscr{Q}}\|_{L^1(\eta_0)}. \tag{7.8}$$

To see that this is so, write

$$\int |\log \rho_{\mathscr{R}} - \log \rho_{\mathscr{Q}}| \, d\eta_0 = \sum_{R \subset Q} \int_R |\log \rho_{\mathscr{R}} - \log \rho_{\mathscr{Q}}| \, d\eta_0,$$

where the sum is over the pairs of atoms $R \in \mathscr{R}$ and $Q \in \mathscr{Q}$ with $R \subset Q$. Since $\rho_{\mathscr{R}}$ and $\rho_{\mathscr{Q}}$ are constant on each $R \in \mathscr{R}$, this may be rewritten as

$$\sum_{R \subset Q} \eta_0(R) |\log \rho_{\mathscr{R}} - \log \rho_{\mathscr{Q}}| = \sum_{R \subset Q} \left| \int_R \log \rho \, d\eta_0 - \eta_0(R) \log \rho_{\mathscr{Q}} \right|$$
$$\leq \sum_{R \subset Q} \int_R |\log \rho - \log \rho_Q| \, d\eta_0 .$$

The combination of these two relations proves (7.8).

We are ready to prove part (2) of the lemma. Let $(\mathscr{Q}_n)_n$ be any refining sequence of partitions with $m(\partial \mathscr{Q}_n) = 0$ and $\operatorname{diam} \mathscr{Q}_n \to 0$. By the martingale convergence theorem (Theorem 5.21), $\rho_{\mathscr{Q}_n} \to \rho$ at η_0-almost every point. By uniform integrability (7.7), it follows that $\log \rho_{\mathscr{Q}_n} \to \log \rho$ in $L^1(\eta_0)$. Given $\varepsilon > 0$, fix n sufficiently large so that

$$\| \log \rho_{Q_n} - \log \rho \|_{L^1(\eta_0)} < \varepsilon/4. \tag{7.9}$$

Given any partition $\mathscr{P}$ with $\eta_0(\partial \mathscr{P}) = 0$, let $\mathscr{R} = \mathscr{P} \vee \mathscr{Q}_n$ (the coarsest partition that refines both $\mathscr{P}$ and $\mathscr{Q}_n$). Combining (7.8) with (7.9), we get

$$\| \log \rho_{\mathscr{R}} - \log \rho \|_{L^1(\eta_0)} < \varepsilon/2. \tag{7.10}$$

Now, let Y be the set of points $x \in X$ such $\mathscr{P}(x) \not\subset \mathscr{Q}_n(x)$. Since $\eta_0(\partial \mathscr{Q}_n) = 0$ we may find $\delta > 0$ such that $\operatorname{diam} \mathscr{P} < \delta$ implies $\eta_0(Y)$ is small enough that $-\eta_0(Y)(\log \eta_0(Y) + \log c) < \varepsilon/4$. Then, combining (7.8) with the fact that $\mathscr{R}(x) = \mathscr{P}(x)$ for all $x \in X \setminus Y$, we find

$$\| \log \rho_{\mathscr{P}} - \log \rho_{\mathscr{R}} \|_{L^1(\eta_0)} \leq \int_Y (|\log \rho_{\mathscr{P}}| + |\log \rho_{\mathscr{R}}|) \, d\eta_0 < \varepsilon/2.$$

Together with (7.10), this implies $\| \log \rho_{\mathscr{P}} - \log \rho \|_{L^1(\eta_0)} < \varepsilon$, as claimed. □

Proof of Lemma 7.6 First, apply Lemma 7.8 with $X = \mathbb{P}\mathbb{R}^d$ and $c = \varepsilon$ and $\mathscr{P} = \mathscr{P}_{n,k}(x)$ and $\eta_0 = m_x$ and $\eta_1 = A(x)_*^{-1} m_{f(x)} + \varepsilon m_x$. Note that

$$\rho = J + \varepsilon = J_\varepsilon \quad \text{and} \quad \rho_{\mathscr{P}} = J_{n,k} + \varepsilon = J_{n,k,\varepsilon}.$$

Moreover, Lemma 7.7 ensures that the diameter of $\mathscr{P}$ is small if n is large enough. It follows from Lemma 7.8 that, for any $\gamma > 0$ and μ-almost every x, one has

$$\| \log J_{n,k,\varepsilon} - \log J_\varepsilon \|_{L^1(m_x)} < \gamma \quad \text{for all } 0 \leq k < n \tag{7.11}$$

if n is sufficiently large n. Note that $\| \log J_\varepsilon \|_{L^1(m_x)} \leq 1 - \log \varepsilon$, because

$$\int_{\{J_\varepsilon \geq 1\}} \log J_\varepsilon \, dm_x \leq 1 \quad \text{and} \quad \int_{\{J_\varepsilon < 1\}} -\log J_\varepsilon \, dm_x \leq -\log \varepsilon,$$

and analogously for $J_{n,k,\varepsilon}$. This shows that the left-hand side of (7.11) is uniformly bounded, and so we may use the bounded convergence theorem to deduce that

$$\|\log J_{n,k,\varepsilon} - \log J_\varepsilon\|_{L^1(m)} < 2\gamma \quad \text{for all } 0 \le k < n$$

if n is sufficiently large. That finishes the proof of the lemma. □

This proves Proposition 7.4 in the ergodic case. Now let m be an arbitrary F-invariant probability measure that projects down to μ, and let $m = \int m_\alpha \, d\alpha$ be an ergodic decomposition of m. Using Exercise 7.3, it follows that the claim of Proposition 7.4 holds for m as well:

$$h(m) = \int \int -\log J \, dm_\alpha \, d\alpha = \int h(m_\alpha) \, d\alpha \le d\Delta.$$

This completes the proof of Proposition 7.4 and Theorem 7.2.

7.3 Furstenberg's criterion

Let us go back to the setting at the beginning of Section 7.1. Let $\nu = \hat{A}_* \hat{\mu}$. Among the consequences of the invariance principle is the following result of Furstenberg [54] (the SL(2)-version appeared already in Exercise 6.10):

Theorem 7.9 (Furstenberg) *If $\lambda_- = \lambda_+$ then there exists some probability measure η on $\mathbb{P}\mathbb{R}^d$ such that $B_*\eta = \eta$ for every $B \in \operatorname{supp} \nu$. Indeed, this holds for any $\mathbb{P}\hat{F}$-stationary measure η.*

Proof By Proposition 5.6, the set of stationary measures is non-empty. Let η be any stationary measure in $\mathbb{P}\mathbb{R}^d$ and let m be its lift to $M \times \mathbb{P}\mathbb{R}^d$. Then m is an $\mathbb{P}F$-invariant probability measure. By Theorem 7.1, m is an *su*-state and so (Exercise 5.17) $m = \mu \times \eta$. Now the fact that m is $\mathbb{P}\hat{F}$-invariant means (Exercise 5.20) that $\hat{A}(x)_*\eta = \eta$ for μ-almost every $x \in M$. □

Example 7.10 Consider the locally constant cocycle defined by

$$A_1 = \begin{pmatrix} 2 & 0 \\ 0 & 2^{-1} \end{pmatrix} \quad \text{and} \quad A_2 = R_\theta \text{ with } \theta \in \mathbb{R} \setminus \mathbb{Q}.$$

On the one hand, the A_1-invariant probability measures in $\mathbb{P}\mathbb{R}^2$ are the convex combinations of the Dirac masses at $[(1,0)]$ and $[(0,1)]$. On the other hand, A_2 has a unique invariant probability measure; namely, the Haar measure in $\mathbb{P}\mathbb{R}^2$. Thus, the two matrices have no common invariant probability measure. Hence, Theorem 7.9 implies that $\lambda_- < 0 < \lambda_+$.

Let us comment a bit on the relations between Theorems 6.11 and 7.9. The former was stated for non-invertible cocycles, whereas presently we deal with the invertible case. However, that distinction is clearly irrelevant here: as we have observed before (for instance, while proving Proposition 6.7), every locally constant cocycle has a trivial extension to an invertible locally constant cocycle. Observe also that any compact subgroup of $\mathrm{GL}(d)$ must be contained in $\mathrm{SL}(d)$, and so the two groups have the same compact subgroups.

It is not difficult to see that, up to these simple observations, Theorem 7.9 contains the high-dimensional version of Theorem 6.11:

Corollary 7.11 (Furstenberg) *Assume that*

(a) *the support of* ν *is not contained in a compact subgroup of* $\mathrm{GL}(d)$*;*
(b) *the cocycle* $\hat{F}$ *is strongly irreducible.*

Then $\lambda_- < \lambda_+$.

Proof According to Exercise 7.5, the conditions (a) and (b) imply that there exists no probability measure η on $\mathbb{P}\mathbb{R}^d$ that is invariant under every $B \in \operatorname{supp} \nu$. The conclusion then follows from Theorem 7.9. □

For $\mathrm{SL}(2)$-cocycles, Theorem 6.11 also contains Theorem 7.9, as we saw in Exercise 6.10. However, the analogue of Exercise 6.10 in dimensions $d \geq 3$ (the converse to Exercise 7.5) is only partly true, by Exercises 7.6 and 7.7.

7.4 Lyapunov exponents of typical cocycles

Let $X = \{1, \ldots, m\}$ be a finite set with $m \geq 2$ elements and let p be a probability measure on X with $p(\{i\}) > 0$ for every $i \in X$. Let $f : M \to M$ be the shift map on $M = p^{\mathbb{N}}$ and consider $\mu = p^{\mathbb{N}}$. The set of functions $A : X \to \mathrm{GL}(d)$ may be identified with $\mathrm{GL}(d)^m$ which, in turn, may be viewed as an open subset of a Euclidean space. Let $\lambda_{\pm}(A)$ denote the extremal Lyapunov exponents of the locally constant linear cocycle $F : M \times \mathbb{R}^d \to M \times \mathbb{R}^d$ defined by A over (f, μ).

Theorem 7.12 *The subset* $\mathscr{Z}$ *of the functions* $A : X \to \mathrm{GL}(d)$ *such that* $\lambda_-(A) = \lambda_+(A)$ *is contained in a finite union of closed proper submanifolds of* $\mathrm{GL}(d)^m$*. In particular, the closure of* $\mathscr{Z}$ *is nowhere dense and has volume zero.*

We prove this theorem later. Let us point out now that the conclusion can be sharpened considerably. Exercise 7.8 is one example of this. Another is that the arguments we are going to present remain valid if one replaces $\mathrm{GL}(d)$ by the subgroup $\mathrm{SL}(d)$: just note that the curves $B(t)$ defined in (7.12) and (7.14) lie in $\mathrm{SL}(d)$ if the initial matrix A does.

7.4.1 Eigenvalues and eigenspaces

For each $r,s \geq 0$ with $r+2s = d$, let $G(r,s)$ be the subset of matrices $A \in \mathrm{GL}(d)$ having r real eigenvalues and s pairs of (strictly) complex conjugate eigenvalues, such that all the eigenvalues that do not belong to the same complex conjugate pair have distinct norms. Every $G(r,s)$ is open and the complement of the union $G = \bigcup_{r,s} G(r,s)$ is a *small subset* of $\mathrm{GL}(d)$, meaning that it is contained in a finite union of closed proper submanifolds.

For $A \in G(r,s)$, let $\lambda_1(A),\ldots,\lambda_r(A)$ be the real eigenvalues, in decreasing order of the absolute value, and $\mu_1(A),\bar{\mu}_1(A),\ldots,\mu_s(A),\bar{\mu}_s(A)$ be the pairs of complex eigenvalues, also in decreasing order of the absolute value. Moreover, let $\xi_j(A) \in \mathrm{Gr}(1,d)$ be the eigenspace corresponding to each real eigenvalue $\lambda_j(A)$ and $\eta_k(A) \in \mathrm{Gr}(2,d)$ be the invariant plane corresponding to each pair of complex conjugate eigenvalues $\mu_k(A)$ and $\bar{\mu}_k(A)$.

We need to analyse how these eigenvalues and invariant subspaces vary inside $G(r,s)$. Let us start with the case when all eigenvalues are real:

Lemma 7.13 *The maps $A \mapsto \lambda_j(A)$ and $A \mapsto \xi_j(A)$ are smooth on $G(d,0)$ for $j = 1,\ldots,d$. Moreover, the map $G(d,0) \to \mathrm{Gr}(1,d)^d$, $A \mapsto (\xi_1(B),\ldots,\xi_d(B))$ is a submersion.*

Proof Each $\lambda_j(A)$ is a simple root of the polynomial $\det(A - \lambda\,\mathrm{id})$, and so it has a smooth continuation on $G(d,0)$, given by the implicit function theorem. Denote $L_j(A) = A - \lambda_j(A)\,\mathrm{id}$. This matrix depends smoothly on $A \in G(d,0)$ and, since $\lambda_j(A)$ remains a simple eigenvalue throughout, it always has rank $d-1$. Let $L_j(A)^a$ be the adjoint matrix. Its entries are the cofactors of $L_j(A)$, and so the adjoint is non-zero and varies smoothly with A. In particular, the columns of $L_j(A)^a$ are smooth functions of A. Moreover, $L_j(A) \cdot L_j(A)^a = \det L_j(A)\,\mathrm{id} = 0$, which implies that every non-zero column of $L_j(A)^a$ is an eigenvector for A, associated with the eigenvalue $\lambda_j(A)$. This shows that every $A \mapsto \xi_j(A)$ is smooth. To check that the derivative of

$$\xi : A \mapsto (\xi_1(A),\ldots,\xi_d(A)) \in \mathrm{Gr}(1,d)^d$$

is onto, consider any differentiable curve $\beta(t) = (\beta_1(t),\ldots,\beta_d(t))$ on $\mathrm{Gr}(1,d)^d$ such that $\beta(0) = \xi(A)$. Take $t \mapsto P(t)$ to be a differentiable curve in $\mathrm{GL}(d)$ such that the columns of $P(t)$ are non-zero vectors in the direction of the $\beta_j(t)$. Define

$$B(t) = P(t)\,\mathrm{diag}[\lambda_1(A),\ldots,\lambda_d(A)]P(t)^{-1}. \quad (7.12)$$

Then $t \mapsto B(t)$ is also a differentiable curve in $\mathrm{GL}(d)$ with $B(0) = A$ and $\xi(B(t)) = \beta(t)$ for all t. In particular, the derivative $D\xi(A)$ maps $B'(0)$ to $\beta'(0)$. So the derivative of ξ is surjective, as claimed. $\square$

Next we deal with complex eigenvalues. Given $A \in \mathrm{GL}(d)$, let μ be a complex solution of $\det(A - \mu\,\mathrm{id}) = 0$ and $v \in \mathbb{C}^d \setminus \{0\}$ be such that $Av = \mu v$. Then, since the matrix A is taken to be real, $\det(A - \bar{\mu}\,\mathrm{id}) = 0$ and $A\bar{v} = \bar{\mu}\bar{v}$. Assuming that $\mu \notin \mathbb{R}$, the conjugate vectors v and $\bar{v}$ are linearly independent and so (Exercise 7.9) the real vectors $\Re v$ and $\Im v$ are also linearly independent. Moreover, the real plane $\psi(v) = \mathrm{span}\{\Re v, \Im v\}$ is invariant under A. This plane depends only on the complex projective class of v and so we may consider

$$\psi : \{[v] \in \mathbb{PC}^d : v \text{ and } \bar{v} \text{ are linearly independent}\} \to \mathrm{Gr}(2,d). \tag{7.13}$$

We invite the reader to check that this map is smooth and is a submersion (Exercise 7.10).

Lemma 7.14 *The maps $A \mapsto \lambda_j(A)$ and $A \mapsto \mu_k(A)$ and $A \mapsto \xi_j(A)$ and $A \mapsto \eta_k(A)$ are smooth on $G(r,s)$, for $j = 1, \ldots, r$ and $k = 1, \ldots s$. Furthermore,*

$$G(r,s) \to \mathrm{Gr}(1,d)^r \times \mathrm{Gr}(2,d)^s, \quad A \mapsto (\xi_1(A), \ldots, \xi_r(A), \eta_1(A), \ldots, \eta_s(A))$$

is a submersion.

Proof Smoothness of the eigenvalues λ_j and μ_k follows from the implicit function theorem. Recall that $\xi_j(A) \in \mathbb{PR}^d$ denotes the eigenspace associated with each real eigenvalue $\lambda_j(A)$. Let $\zeta_k(A) \in \mathbb{PC}^d$ be the eigenspace associated with each complex eigenvalue $\mu_k(A)$. The same arguments as in Lemma 7.13 imply that $A \mapsto \xi_j(A)$ and $A \mapsto \zeta_k(A)$ are smooth maps. By the observations preceding this lemma, it follows that $\eta_k(A) = \psi(\zeta_k(A)) \in \mathrm{Gr}(2,d)$ is also a smooth function of A. Moreover (Exercise 7.10), to prove that the map in the statement is a submersion it suffices to show that

$$\zeta : A \mapsto (\xi_1(A), \ldots, \xi_r(A), \zeta_1(A), \ldots, \zeta_s(A)) \in (\mathbb{PR}^d)^r \times (\mathbb{PC}^d)^s$$

is a submersion. Let $\alpha(t) = (\beta_1(t), \ldots, \beta_r(t), \gamma_1(t), \ldots, \gamma_s(t))$ be any differentiable curve in $(\mathbb{PR}^d)^r \times (\mathbb{PC}^d)^s$ with $\alpha(0) = \zeta(A)$. Define

$$\begin{aligned} B(t) = P(t)\,\mathrm{diag}[&\lambda_1(A), \ldots, \lambda_r(A), \\ &\mu_1(A), \bar{\mu}_1(A), \ldots, \mu_s(A), \bar{\mu}_s(A)]\,P(t)^{-1}. \end{aligned} \tag{7.14}$$

where $t \mapsto P(t)$ is a differentiable curve such that the columns of $P(t)$ are non-zero vectors $u_j(t) \in \beta_j(t)$ and $v_k(t), \bar{v}_k(t)$ with $v_k(t) \in \gamma_k(t)$. Observe that $t \mapsto B(t)$ is a differentiable curve in $\mathrm{GL}(d)$, even if $P(t)$ need not be real. Moreover, $B(0) = A$ and $\zeta(B(t)) = \alpha(t)$ for all t. So $D\zeta(A)$ maps $B'(0)$ to $\alpha'(0)$. This shows that the derivative $D\zeta(A)$ is indeed surjective. $\square$

7.4.2 Proof of Theorem 7.12

Since the complement of G is a small subset of $\mathrm{GL}(d)$, the set of $A \in \mathrm{GL}(d)^m$ such that $A_i \notin G$ for some $i = 1, \dots, m$ is a small subset of $\mathrm{GL}(d)^m$. So, we may restrict ourselves to the open set

$$\mathscr{G} = \{A \in \mathrm{GL}(d)^m : A_i \in G \text{ for every } i = 1, \dots, m\}.$$

In view of Theorem 7.9, it suffices to show that the subset of $A \in \mathscr{G}$ such that the matrices A_i, $i = 1, \dots, m$ admit a common invariant probability is small.

If $A_i \in G(r_i, s_i)$ then, as all the eigenvalues have different absolute values, every A_i-invariant probability measure on $\mathbb{PR}^d$ is a convex combination of Dirac masses on the eigenspaces $\xi_j(A)$, $j = 1, \dots, r_i$ and of probability measures supported in the invariant planes $\eta_k(A)$, $k = 1, \dots, s_i$ (each $\eta_k(A)$ is naturally identified with a 1-dimensional subspace of the projective space). In particular, the support must be contained in

$$\Sigma(A_i) = \{\xi_1(A_i), \dots, \xi_{r_i}(A_i)\} \cup \eta_1(A_i) \cup \dots \cup \eta_{s_i}(A_i).$$

Suppose first that $d \geq 4$. Since the $\xi_j(A_i)$ are points and the $\eta_k(A_i)$ are lines in the projective space, and $\dim \mathbb{PR}^d \geq 3$, it follows from Lemmas 7.13 and 7.14 that there exists a small subset $\mathscr{Z}_1$ of $G(r_1, s_1) \times G(r_2, s_2) \times \mathrm{GL}(d)^{m-2}$ such that

$$\xi_a(A_1) \neq \xi_b(A_2) \tag{7.15}$$

$$\xi_a(A_1) \notin \eta_e(A_2) \quad \text{and} \quad \xi_b(A_2) \notin \eta_c(A_1) \tag{7.16}$$

$$\eta_c(A_1) \cap \eta_e(A_2) = \emptyset \tag{7.17}$$

for every $A \in G(r_1, s_1) \times G(r_2, s_2) \times \mathrm{GL}(d)^{m-2} \setminus \mathscr{Z}_1$ and $1 \leq a \leq r_1, 1 \leq b \leq r_2$, $1 \leq c \leq s_1$ and $1 \leq e \leq s_2$. This implies that $\Sigma(A_1) \cap \Sigma(A_2) = \emptyset$ which, by the remarks in the previous paragraph, implies that A_1 and A_2 have no common invariant probability measure. This proves the theorem in dimension $d \geq 4$.

Now suppose that $d = 3$. The previous arguments extend to this case when either $s_1 = 0$ or $s_2 = 0$; that is, when the condition (7.17) is void. When $s_1 = s_2 = 1$, there is one difficulty: since the projective space $\mathbb{PR}^3$ is only 2-dimensional, one cannot force the pair of 1-dimensional submanifolds $\eta_1(A_1)$ and $\eta_1(A_2)$ to be disjoint, as required in (7.17). This can be bypassed as follows.

By the same arguments as before, there exists a small subset of $\mathscr{Z}_2$ of $G(1,1)^2 \times \mathrm{GL}(d)^{m-2}$ such that every $A \in G(1,1) \times G(2,2) \times \mathrm{GL}(d)^{m-2} \setminus \mathscr{Z}_2$ satisfies (7.15) and (7.16) and $\eta_1(A_1) \neq \eta_1(A_2)$ (instead of (7.17)). Suppose that A_1 and A_2 have some common invariant probability measure θ. Observing that

$$\Sigma(A_1) \cap \Sigma(A_2) = \eta_1(A_1) \cap \eta_1(A_2)$$

consists of a single point z in the projective space, θ has to be the Dirac mass at z. Then z has to be a fixed point for A_i inside $\eta_1(A_i)$, for $i = 1,2$. This is impossible, because the eigenspace $\eta_i(A_i)$ contains no A_i-invariant line. This contradiction proves the theorem in dimension $d = 3$.

Now we treat the case $d = 2$. If $s_1 = s_2 = 0$ then the conditions (7.16) and (7.17) are void and we can use the same arguments as we did for $d \geq 4$. If $s_1 = 0$ and $s_2 = 1$ then $\Sigma(A_1) \cap \Sigma(A_2)$ consists of the two points $\xi_1(A_1)$ and $\xi_2(A_1)$. So, if A_1 and A_2 have a common invariant probability, then this must be a convex combination of the Dirac masses at the two eigenspaces of A_1. Then the union of these eigenspaces has to be invariant under A_2, which is impossible because the action of $A_2 \in G(0,1)$ on the projective space is a rotation whose angle is *not* a multiple of π. Up to exchanging the roles of the matrices, this also covers the case $d = 2$ with $s_1 = 1$ and $s_2 = 0$.

The only remaining case, $d = 2$ with $s_1 = s_2 = 1$, requires an essentially new ingredient that we are going to borrow from Douady and Earle [50]. Recall that every matrix $B \in \mathrm{GL}(2)$ with positive determinant induces an automorphism h_B of the Poincaré half-plane $\mathbb{H}$:

$$B = \begin{pmatrix} a & b \\ c & d \end{pmatrix} \quad \longrightarrow \quad h_B(z) = \frac{az+b}{cz+d}. \tag{7.18}$$

The action of B on the projective plane may be identified with the action of h_B on the boundary of $\mathbb{H}$, via

$$\partial\mathbb{H} \to \mathbb{PR}^2, \qquad x \mapsto [(x,1)]$$

(including $x = \infty$) so that B-invariant measures on the projective plane may be seen as h_B-invariant measures sitting on the real axis. It is also easy to check that h_B has a fixed point in the open disc $\mathbb{H}$ if and only if $B \in G(0,1)$. Define $\phi(B)$ to be this (unique) fixed point. It is easy to see that $B \mapsto \phi(B)$ is a smooth submersion: use the explicit expression for the fixed point extracted from (7.18).

Lemma 7.15 *If A_1, $a_2 \in G(0,1)$ have a common invariant probability measure μ on $\partial\mathbb{H}$ then $\phi(A_1) = \phi(A_2)$.*

Proof It is clear that A_1 and A_2 have no invariant measures with atoms of mass larger than $1/3$: such atoms would correspond to periodic points in the projective plane, with periods 1 or 2, which are forbidden by the definition of $G(0,1)$. In Proposition 1 of [50] a map $\mu \mapsto B(\mu)$ is constructed that assigns to each probability measure μ with no atoms of mass $\geq 1/2$ (see Remark 2 in [50, page 26]) a point $B(\mu)$ in the half-plane $\mathbb{H}$, called *conformal barycenter* of μ, such that

$$B(h_*\mu) = h(B(\mu)) \quad \text{for every conformal automorphism } h : \mathbb{H} \to \mathbb{H}.$$

When μ is A_i-invariant this implies $h_{A_i}(B(\mu)) = B((h_{A_i})_*\mu) = B(\mu)$, and so the conformal barycenter must coincide with the fixed point $\phi(A_i)$ of the automorphism h_{A_i}. Thus, if μ is a invariant measure for both A_1 and A_2 then $\phi(A_1) = B(\mu) = \phi(A_2)$. □

Since ϕ is a submersion, $\mathscr{Z}_3 = \{A \in G(0,1)^2 \times \mathrm{GL}(d)^{m-2} : \phi(A_1) = \phi(A_2)\}$ is a small subset of $G(0,1)^2 \times \mathrm{GL}(d)^{m-2}$. By Lemma 7.15, if A is in the complement of $\mathscr{Z}_3$ then there are no common invariant probability measures.

Altogether, this proves that $\mathscr{Z}$ is a small subset of $\mathrm{GL}(d)^m$, as claimed in the first part of Theorem 7.12. The second part is an immediate consequence, because the closure of a small subset is still a small subset, and small subsets have volume zero and, consequently, are nowhere dense.

7.5 Notes

Theorem 7.2 is a special case of the main result of Ledrappier [81]. Ledrappier shows how to deduce both Furstenberg's theorem (Theorem 7.9) and a celebrated theorem of Kotani [77]. The proof in Section 7.2 is from Avila and Viana [16], where a much more general statement is obtained.

The invariance principle for linear cocycles goes back to Bonatti, Gomez-Mont and Viana [37]. They considered both locally constant cocycles and Hölder-continuous cocycles with invariant holonomies over hyperbolic homeomorphisms with invariant probabilities having local product structure (the notion of local product structure was defined in the Notes to Chapter 5 and that of invariant holonomies will be discussed in Section 10.6). Viana [113] used a non-uniform version of these methods to prove that almost all linear cocycles over any hyperbolic system, uniform or not, have some non-vanishing Lyapunov exponent.

The expression *invariance principle* was coined by Avila and Viana [16], who extended the statement to smooth cocycles over hyperbolic homeomorphisms; that is, cocycles that act by diffeomorphisms on the fibers. Avila, Santamaria and Viana [13] further extended the statement to smooth cocycles over partially hyperbolic volume-preserving diffeomorphisms. This turned the invariance principle into a very flexible tool, with several applications in the theory of partially hyperbolic systems.

Among these, we mention the rigidity results of Avila and Viana [16], for symplectic diffeomorphisms, and Avila, Viana and Wilkinson [17], for time-one maps of geodesic flows, the construction of measures of maximal entropy by Rodriguez-Hertz, Rodriguez-Hertz, Tahzibi and Ures [64] and the results of Viana and Yang [115] on the existence and finiteness of physical measures.

Theorem 7.9 and Corollary 7.11 extend Theorem 6.11 to arbitrary dimension. They were originally due to Furstenberg [54] and also hold more generally than stated (see [81, Corollary 1]): it suffices to assume that $\mathscr{A}_+ \cap \mathscr{A}_-$ contains only zero and full measure sets, where $\mathscr{A}_\pm$ is the σ-algebra generated by the functions $A \circ f^{\pm n}$, $n \geq 1$. Clearly, this is the case if (f, μ) is a Bernoulli shift and A is locally constant.

Section 7.4 is a simplified version of material from Bonatti, Gomez-Mont and Viana [37], which has also been extended by Avila, Santamaria and Viana [13] to cocycles over partially hyperbolic diffeomorphisms. In fact, for the 2-dimensional case we follow the latter paper more closely.

7.6 Exercises

Exercise 7.1 Let $N = \mathbb{PR}^d$ and $\Psi : N \to N$ be the projective action of some $B \in \mathrm{GL}(d)$. Prove that, for every $[v] \in N$ and $\xi \in T_{[v]}N$,

(1) $T_{[v]}N =$ orthogonal complement of $[v]$;

(2) $D\Psi([v])\xi =$ orthogonal projection of $\dfrac{\|v\|}{\|B(v)\|}B(\xi)$ to $T_{[B(v)]}N$;

(3) $\|D\Phi([v])\| \leq \|B\|\|B^{-1}\|$ and $\|D\Phi([v])^{-1}\| \leq \|B\|\|B^{-1}\|$;

(4) $|\det D\Psi([v])| = |\det B| \left(\dfrac{\|v\|}{\|B(v)\|}\right)^d$.

Exercise 7.2 Let θ be a probability measure on $\mathbb{PR}^d$ and consider any finite family of projective hyperplanes $H_1, \ldots, H_n \subset \mathbb{PR}^d$. Show that there exists some orthogonal transformation $B : \mathbb{R}^d \to \mathbb{R}^d$ such that $\theta(B(H_j)) = 0$ for every $j = 1, \ldots, n$.

Exercise 7.3 Prove that if m is an F-invariant probability measure that projects down to μ then almost every ergodic component m_α projects down to μ.

Exercise 7.4 Prove that Corollary 6.15 extends to any dimension $d \geq 2$. Namely, that the *stabilizer* $H(\theta) = \{B \in \mathrm{GL}(d) : B_*\theta = \theta\}$ of any non-atomic probability measure θ on $\mathbb{PR}^d$ is a compact subgroup of $\mathrm{GL}(d)$.

Exercise 7.5 Show that Exercise 6.9 extends to any dimension $d \geq 2$. Namely, that if (a) the support of ν is contained in no compact subgroup of $\mathrm{GL}(d)$ and (b) the cocycle is strongly irreducible, then there exists no probability measure on $\mathbb{PR}^d$ such that $B_*\eta = \eta$ for every $B \in \operatorname{supp}\nu$.

Exercise 7.6 Show that if the support of ν is contained in a compact subgroup of $\mathrm{GL}(d)$, $d \geq 2$ then there exists some probability measure on $\mathbb{P}\mathbb{R}^d$ such that $B_*\eta = \eta$ for every $B \in \operatorname{supp} \nu$.

Exercise 7.7 Show that there exist $\mathrm{SL}(3)$-cocycles that are not irreducible and yet admit no probability measure on $\mathbb{P}\mathbb{R}^3$ with $B_*\eta = \eta$ for every $B \in \operatorname{supp} \nu$. Check that this is the case for any random product of $A_1 =$ shear along the xy-plane, $A_2 =$ irrational rotation around the z-axis and $A_3 =$ hyperbolic map on the xy-plane cross identity along the z-axis, with probabilities $p_1, p_2, p_3 > 0$.

Exercise 7.8 Check that the submanifolds in Theorem 7.12 may be chosen with codimension $\geq [m/2]$.

Exercise 7.9 Given $v \in \mathbb{C}^d$, prove that the following conditions are equivalent:

(1) v and $\bar{v}$ are $\mathbb{C}$-linearly dependent;
(2) $\Re v$ and $\Im v$ are $\mathbb{R}$-linearly dependent;
(3) $v = e^{i\theta} w$ for some $\theta \in \mathbb{R}$ and $w \in \mathbb{R}^d$.

Exercise 7.10 Show that the set $\mathscr{R} = \{[w] \in \mathbb{P}\mathbb{C}^d : w \in \mathbb{C}^d \setminus \{0\}\}$ is closed in $\mathbb{P}\mathbb{C}^d$ and the map $\psi : \mathbb{P}\mathbb{C}^d \setminus \mathscr{R} \to \mathrm{Gr}(2,d)$ in (7.13) is a submersion. This can be done in the following steps.

(1) Let $1 \leq i < j \leq d$ be fixed. Given vectors $u, v \in \mathbb{R}^d$, let $\pi_{i,j}(u,v)$ be the 2×2-matrix whose rows are (u_i, v_i) and (u_j, v_j), and $\pi^*_{i,j}(u,v)$ be the $(d-2) \times 2$-matrix whose rows are (u_l, v_l) for $l \notin \{i,j\}$. Denote by $G_{i,j}$ the subset of planes $L \in \mathrm{Gr}(2,d)$ admitting a basis $\{u,v\}$ such that $\pi_{i,j}(u,v)$ is invertible. Check that the matrix $\phi_{i,j}(L) = \pi^*_{i,j}(u,v)\pi_{i,j}(u,v)^{-1}$ does not depend on the choice of the basis, and these maps $\phi_{i,j} : G_{i,j} \to \mathbb{R}^{2(d-2)}$ form an atlas of $\mathrm{Gr}(2,d)$.
(2) Observe that the map ψ is given by $\psi([w]) = \pi^*_{i,j}(\Re w, \Im w)\pi_{i,j}(\Re w, \Im w)^{-1}$ in such local coordinates. Deduce that ψ is smooth and the derivative is given by

$$\begin{aligned} D\psi([w])\dot{w} &= \pi^*_{i,j}(\Re \dot{w}, \Im \dot{w})\pi_{i,j}(\Re w, \Im w)^{-1} \\ &\quad - \pi^*_{i,j}(\Re w, \Im w)\pi_{i,j}(\Re w, \Im w)^{-1}\pi_{i,j}(\Re \dot{w}, \Im \dot{w})\pi_{i,j}(\Re w, \Im w)^{-1}, \end{aligned}$$

for every $\dot{w} \in T_{[w]}\mathbb{P}\mathbb{C}^d$. Given $\dot{B} \in T_{\psi([w])}\mathrm{Gr}(2,d)$ (a real $(d-2) \times 2$ matrix), consider $\dot{u}, \dot{v} \in \mathbb{R}^d$ such that $\pi^*_{i,j}(\dot{u}, \dot{v}) = \dot{B}\pi_{i,j}(\Re w, \Im w)$ and $\pi_{i,j}(\dot{u}, \dot{v}) = 0$. Let $\dot{w} = \dot{u} + i\dot{v}$ and check that $D\psi([w])\dot{w} = \dot{B}$. Conclude that the derivative of ψ is indeed surjective.

8
Simplicity

In the previous two chapters we found sufficient conditions for the largest and smallest Lyapunov exponents λ_- and λ_+ to be distinct. We are now going to see that, under only mildly stronger conditions, *all* Lyapunov exponents are distinct.

Let $F : M \times \mathbb{R}^d \to M \times \mathbb{R}^d$ be the linear cocycle defined over a measure-preserving transformation $f : (M,\mu) \to (M,\mu)$ by some measurable function $A : M \to \mathrm{GL}(d)$ such that $\log^+ \|A^{\pm 1}\| \in L^1(\mu)$. The Lyapunov spectrum of F is *simple* if all the Lyapunov exponents have multiplicity 1; that is, if

$$\dim V_x^j - \dim V_x^{j+1} = 1 \quad \text{for every } j, \tag{8.1}$$

where $\mathbb{R}^d = V_x^1 > \cdots > V_x^k > \{0\}$ is the Oseledets flag in Theorem 4.1. In other words, F has d distinct Lyapunov exponents. When F is invertible, condition (8.1) may be written as $\dim E_x^j = 1$ for every j, where $E_x^1 \oplus \cdots \oplus E_x^k$ is the Oseledets decomposition in Theorem 4.2.

The main result in this chapter (Theorem 8.1) is a general criterion for simplicity. We restrict ourselves to the case when $F : M \times \mathbb{R}^d \to M \times \mathbb{R}^d$ is locally constant, although the statement is a lot more general. That is, we suppose that (f,μ) is a Bernoulli shift, either one-sided or two-sided, and the function $A : M \to \mathrm{GL}(d)$ depends only on the zeroth coordinate. Let $\nu = A_*\mu$ be the probability measure on $\mathrm{GL}(d)$ obtained by push-forward of μ under A.

We need to recall a few elementary notions from linear algebra.

8.1 Pinching and twisting

The *singular values* of a linear operator $B \in \mathrm{GL}(d)$ are the positive square roots $\sigma_1, \ldots, \sigma_d$ of the eigenvalues of the self-adjoint operator $B^t B$. They have the following geometric interpretation: the image of the unit sphere under $B : \mathbb{R}^d \to$

$\mathbb{R}^d$ is an ellipsoid with semi-axes of length $\sigma_1, \ldots, \sigma_d$. We always take the singular values to be numbered in non-increasing order: $\sigma_1(B) \geq \cdots \geq \sigma_d(B)$. The *eccentricity* of B is defined by

$$\operatorname{ecc}(B) = \inf_{1 \leq k \leq d-1} \frac{\sigma_k(B)}{\sigma_{k+1}(B)}.$$

Given a $(d-k)$-dimensional subspace G of $\mathbb{R}^d$, the *hyperplane section* dual to G is the set of all $F \in \operatorname{Gr}(k,d)$ such that $F \cap G \neq \{0\}$. This is a closed, nowhere-dense subset of $\operatorname{Gr}(k,d)$.

A *monoid* is a set $\mathscr{B}$ endowed with an associative binary operation that admits a unit element. The monoid associated with the linear cocycle F is the set $\mathscr{B} \subset \operatorname{GL}(d)$ of all products $B_1 \cdots B_n$ with $B_i \in \operatorname{supp} \nu$ for $1 \leq i \leq n$ and $n \geq 0$ (for $n = 0$ interpret the product to be the identity). A monoid $\mathscr{B} \subset \operatorname{GL}(d)$ is

- *pinching* if it contains matrices with arbitrarily large eccentricity;
- *twisting* if, given any $1 \leq k \leq d-1$, any $F \in \operatorname{Gr}(k,d)$, and any finite family $G_1, \ldots, G_n \in \operatorname{Gr}(d-k,d)$, there exists $B \in \mathscr{B}$ such that $B(F) \cap G_i = \{0\}$ for every $1 \leq i \leq n$.

We leave it to the reader to check that for $d = 2$ these conditions coincide with those in the definitions of pinching and twisting in Section 1.2.

Theorem 8.1 *Suppose that the monoid $\mathscr{B}$ associated with F is pinching and twisting. Then the Lyapunov spectrum of the linear cocycle F is simple.*

The hypotheses of the theorem are typical among locally constant cocycles (Exercise 8.1). It is also interesting to compare these hypotheses with those of the high-dimensional Furstenberg theorem. It is clear that pinching implies the non-compactness condition (a) in Corollary 7.11 and the implication is strict if $d \geq 3$. Moreover, twisting implies the strong irreducibility condition (b) in Corollary 7.11. Lemma 8.7 below contains a kind of converse to this last observation.

8.2 Proof of the simplicity criterion

For proving Theorem 8.1, it is no restriction to suppose the Bernoulli shift $f : (M, \mu) \to (M, \mu)$ to be two-sided, $M = X^{\mathbb{Z}}$ and $\mu = p^{\mathbb{Z}}$, and we do so.

Given $B \in \operatorname{GL}(d)$ and $1 \leq k \leq d-1$ such that $\sigma_k(B) > \sigma_{k+1}(B)$, define $E_k^u(B) \in \operatorname{Gr}(k,d)$ and $E_k^s(B) \in \operatorname{Gr}(d-k,d)$ to be the subspaces given by:

- $E_k^u(B)$ is spanned by the eigenvectors of $B^t B$ with eigenvalue $\geq \sigma_k(B)^2$;

- $E_k^s(B)$ is spanned by the eigenvectors of B^tB with eigenvalue $\leq \sigma_{k+1}(B)^2$.

In geometric terms: $E_k^u(B)$ is the pre-image of the subspace generated by the k largest semi-axes, and $E_k^u(B)$ is the pre-image of the subspace generated by the $d-k$ smallest semi-axes of the ellipsoid $\{B(v) : \|v\| = 1\}$.

Proposition 8.2 *For each $1 \leq k \leq d-1$ there exists a measurable function $\xi^u : M^- \to \mathrm{Gr}(k,d)$ such that $\hat{\xi}^u = \xi^u \circ \pi^- : M \to \mathrm{Gr}(k,d)$ satisfies*

(1) $\hat{\xi}^u(f(x)) = A(x)\hat{\xi}^u(x)$ *for μ-almost every $x \in M$.*
(2) *For μ-almost every $x \in M$,*

$$\frac{\sigma_{d-k}(A^{-n}(x))}{\sigma_{d-k+1}(A^{-n}(x))} \to \infty \quad \textit{and} \quad E_{d-k}^s(A^{-n}(x)) \to \hat{\xi}^u(x).$$

(3) *For every hyperplane section $S \subset \mathrm{Gr}(k,d)$ there exists a set of $x^- \in M^-$ with positive μ^--measure such that $\xi^u(x^-) \notin S$.*

The proof of this proposition will be given in a while. Replacing k by $d-k$ and reversing time, we also obtain:

Proposition 8.3 *For each $1 \leq k \leq d-1$ there exists a measurable function $\xi^s : M^+ \to \mathrm{Gr}(d-k,d)$ such that $\hat{\xi}^s = \xi^s \circ \pi^+ : M \to \mathrm{Gr}(d-k,d)$ satisfies*

(1) $\hat{\xi}^s(f(x)) = A(x) \cdot \hat{\xi}^s(x)$ *for μ-almost every $x \in M$.*
(2) *For μ–almost every $x \in M$,*

$$\frac{\sigma_k(A^n(x))}{\sigma_{k+1}(A^n(x))} \to \infty \quad \textit{and} \quad E_k^s(A^n(x)) \to \hat{\xi}^s(x).$$

(3) *For every hyperplane section $S \subset \mathrm{Gr}(d-k,d)$ there exists a set of $x^+ \in M^+$ with positive μ^+-measure such that $\xi^s(x^+) \notin S$.*

Proof of Theorem 8.1 Let $1 \leq k \leq d-1$ be fixed. Let $\xi^u : M^- \to \mathrm{Gr}(k,d)$ be as in Proposition 8.2 and $\xi^s : M^+ \to \mathrm{Gr}(d-k,d)$ be as in Proposition 8.3. We write each $x \in M$ as (x^-, x^+) with $x^- = \pi^-(x)$ and $x^+ = \pi^+(x)$. We claim that

$$\xi^u(x^-) \cap \xi^s(x^+) = \{0\} \quad \text{for } \mu\text{–almost every } x \in M. \tag{8.2}$$

Indeed, the set of $x \in M$ for which this property holds is (f,μ)-invariant. So, by ergodicity, either the claim is true or else $\xi^u(x^-) \cap \xi^s(x^+) \neq \{0\}$ for μ-almost every x. By Fubini, the latter would imply that there exists $x^+ \in M^+$ such that $\xi^u(x^-)$ is contained in the hyperplane section dual to $\xi^s(x^+)$ for μ^--almost every $x^- \in M^-$, and that would contradict Proposition 8.2. This contradiction proves (8.2).

Thus, we have an F-invariant measurable decomposition $\mathbb{R}^d = \hat{\xi}^s(x) \oplus \hat{\xi}^u(x)$ defined at almost every point. We want to prove that the Lyapunov exponents of

F along $\hat{\xi}^u$ are strictly bigger than those along $\hat{\xi}^s$. Denote $\xi_n^s(x) = E_k^s(A^n(x))$. By Proposition 8.3, the sequence ξ_n^s converges to $\hat{\xi}^s$ at μ-almost every point. In particular, the angle between $\xi_n^s(x)$ and $\hat{\xi}^u(x)$ is bounded from zero for all large n. So, we may choose $\alpha > 0$, $n_0 \geq 1$, and a set $Z \subset M$ with $\mu(Z) > 0$ such that

$$\hat{\xi}^u(x) \subset C(\xi_n^s(x)^\perp, \alpha) \quad \text{and} \quad \hat{\xi}^s(x) \cap C(\xi_n^s(x)^\perp, 4\alpha) = \emptyset \tag{8.3}$$

for every $n \geq n_0$ and $x \in Z$. Note that $C(V,a)$ denotes the cone of radius $a > 0$ around a subspace V of $\mathbb{R}^d$, as defined in Section 4.4.2. Fix $\beta \in (0, 1/10)$ such that (Exercise 4.10)

$$C(\hat{\xi}^u(x), 4\beta) \subset C(\xi_n^s(x)^\perp, 2\alpha).$$

Then, for any $x \in Z$ and $n \geq n_0$,

$$A^n(x)\big(C(\hat{\xi}^u(x), 4\beta)\big) \subset C\big(A^n(x)(\xi_n^s(x)^\perp), 2\alpha\mathscr{E}_n(x)\big) \tag{8.4}$$

where

$$\mathscr{E}_n(x) = \frac{\sigma_{k+1}(A^n(x))}{\sigma_k(A^n(x))}.$$

By Proposition 8.3, $\mathscr{E}_n(x)$ goes to zero as $n \to \infty$. Thus, increasing n_0 and reducing Z if necessary, we may suppose $2\alpha\mathscr{E}_n(x) \leq \beta$ for every $n \geq n_0$ and $x \in Z$. Now, (8.4) and the fact that $\hat{\xi}^u$ is F-invariant imply that

$$\hat{\xi}^u(f^n(x)) \in C\big(A^n(x)(\xi_n^s(x)^\perp), \beta\big).$$

Then $C\big(A^n(x)(\xi_n^s(x)^\perp), \beta\big) \subset C\big(\hat{\xi}^u(f^n(x)), 3\beta\big)$ (Exercise 4.10). Altogether, this proves that

$$A^n(x)\big(C(\hat{\xi}^u(x), 4\beta)\big) \subset C\big(\hat{\xi}^u(f^n(x)), 3\beta\big) \tag{8.5}$$

for every $n \geq n_0$ and $x \in Z$.

Reducing Z once more, if necessary, we may assume that its first n_0 iterates are pairwise disjoint. Equivalently, the first return time $r(x) \geq n_0$ for every $x \in Z$. Let $g(x) : Z \to Z$, $g(x) = f^{r(x)}(x)$ be the first return map and $G : Z \times \mathbb{R}^d \to Z \times \mathbb{R}^d$ be the linear cocycle $G(x,v) = (g(x), B(x)v)$ with $B(x) = A^{r(x)}(x)$. Then property (8.5) implies that every $B(x)$ maps the cone $C(\hat{\xi}^u(x), 4\beta)$ inside a strictly smaller cone $C(\hat{\xi}^u(f^n(x)), 3\beta)$. By (8.3), these cones are disjoint from $\hat{\xi}^s(x)$. So, we are in a position to apply Proposition 4.19 to conclude that the Lyapunov exponents of G along $\hat{\xi}^u(x)$ are strictly bigger than the Lyapunov exponents of G along $\hat{\xi}^s(x)$. Then, by Proposition 4.18, the same is true for the original cocycle F: the first k exponents are strictly larger than all the remaining ones. Since k is arbitrary, this proves that all the Lyapunov exponents of F are distinct, as claimed in the theorem. $\square$

Now let us prove Proposition 8.2 and Proposition 8.3.

8.3 Invariant section

Let $F_k : M \times \mathrm{Gr}(k,d) \to M \times \mathrm{Gr}(k,d)$ be the *Grassmannian cocycle* associated with the linear cocycle $F : M \times \mathbb{R}^d \to M \times \mathbb{R}^d$. That is, $F_k(x,V) = (f(x), A(x)V)$ for every $(x,V) \in M \times \mathrm{Gr}(k,d)$. For $k = 1$ this is just the projective cocycle $\mathbb{P}F : M \times \mathbb{P}\mathbb{R}^d \to M \times \mathbb{P}\mathbb{R}^d$.

For the proofs of Propositions 8.2 and 8.3, we need to expand a bit more on the formalism introduced in Section 4.3.2.

8.3.1 Grassmannian structures

Let $E = \mathbb{R}^d$ and ΛE be the disjoint union of the exterior powers $\Lambda^k E$ with $0 \le k \le d$. The *exterior product* on E is the bilinear map

$$\wedge : \Lambda E \times \Lambda E \to \Lambda E,$$

characterized by $(u_1 \wedge \cdots \wedge u_k) \wedge (v_1 \wedge \cdots \wedge v_l) = (u_1 \wedge \cdots u_k \wedge v_1 \wedge \cdots \wedge v_l)$.

Let H be a hyperplane (that is, a codimension-1 linear subspace) of the vector space $\Lambda^k E$. Then H may be written as

$$H = \{\omega \in \Lambda^k E : \omega \wedge \upsilon = 0\}$$

for some non-zero $\upsilon \in \Lambda^{d-k} E$. We call the hyperplane *geometric* if υ may be chosen to be a decomposable $(d-k)$-vector; that is, $\upsilon = \upsilon_{k+1} \wedge \cdots \wedge \upsilon_d$ for some choice of vectors υ_i in E. Observe that,

$$\omega \in H \Leftrightarrow \omega \wedge \upsilon = 0 \Leftrightarrow \Psi(\omega) \cap \Psi(\upsilon) \neq \{0\} \quad \text{for any } \omega = \omega_1 \wedge \cdots \wedge \omega_k$$

(the map Ψ was introduced in Section 4.3.2). Thus, the intersection of $\mathrm{Gr}(k,d)$ with the projective image of the geometric hyperplane H coincides with the hyperplane section of $\mathrm{Gr}(k,d)$ dual to the $(d-k)$-dimensional subspace $\Psi(\upsilon)$ generated by the vectors $\upsilon_{k+1}, \ldots, \upsilon_d$. This shows that the hyperplane sections of the Grassmannian manifold are, precisely, the intersections of $\mathrm{Gr}(k,d)$ with the projective images of geometric hyperplanes of $\Lambda^k E$. As observed before, hyperplane sections are closed nowhere-dense subsets of the Grassmannian manifold.

A *geometric subspace* of $\Lambda^k E$ is a finite intersection of geometric hyperplanes. Similarly, a *linear section* of $\mathrm{Gr}(k,d)$ is a finite intersection of hyperplane sections. Equivalently, a linear section is the intersection of the Grassmannian with the projective image of some geometric subspace of the exterior power.

We call *linear arrangement* in $\Lambda^k E$ any finite union of geometric subspaces. A *linear arrangement* in $\mathrm{Gr}(k,d)$ is a finite union of linear sections. Thus, a linear arrangement in $\mathrm{Gr}(k,d)$ is the intersection of the Grassmannian manifold with the projective image of some linear arrangement in $\Lambda^k E$. In particular, every linear arrangement is a closed nowhere-dense subset of the Grassmannian.

Lemma 8.4 *If $\{\mathscr{L}_\alpha : \alpha \in I\}$ is an arbitrary family of linear arrangements in* $\mathrm{Gr}(k,d)$ *then $\bigcap_{\alpha\in I} \mathscr{L}_\alpha$ coincides with the intersection of the $\mathscr{L}_\alpha$ over a finite subfamily.*

Proof By definition, for each $\alpha \in I$ there exists some linear arrangement L_α in $\Lambda^k E$ such that $\mathscr{L}_\alpha = \mathbb{P}L_\alpha \cap \mathrm{Gr}(k,d)$. Let $I_1 \subset I$ be an arbitrary finite subset. We may write

$$\bigcap_{\alpha\in I_1} L_\alpha = V_1 \cup \cdots \cup V_m$$

where every $V_j \subset \Lambda^k E$ is a geometric subspace. Renumbering the V_j if necessary, we may suppose that there is $s \in \{0,\ldots,m\}$ such that $V_j \subset \bigcap_{\alpha\in I} \mathscr{L}_\alpha$ if and only if $1 \le j \le s$. If $s = m$ then $\bigcap_{\alpha\in I_1} L_\alpha = \bigcap_{\alpha\in I} L_\alpha$ and so the claim is proved. Otherwise, for each $j \in \{s+1,\ldots,n\}$ there is $\alpha_j \in I$ such that V_j is not contained in L_{α_j}. Let $I_2 = I_1 \cup \{\alpha_{s+1},\ldots,\alpha_n\}$. Then

$$\bigcap_{\alpha\in I_2} L_\alpha = W_1 \cup \cdots \cup W_n$$

where $W_i = V_i$ for $i \le s$ and each W_i with $i > s$ is a proper subspace of some V_j with $j > s$. In particular,

$$\max\Big\{\dim W_j : W_j \not\subset \bigcap_{\alpha\in I} L_\alpha\Big\} \le \max\Big\{\dim V_j : V_j \not\subset \bigcap_{\alpha\in I} L_\alpha\Big\} - 1.$$

Thus, repeating this procedure not more than $\dim \Lambda^k E$ times, we find a finite set $I_r \subset I$ such that $\bigcap_{\alpha\in I_r} L_\alpha = \bigcap_{\alpha\in I} L_\alpha$. Then

$$\bigcap_{\alpha\in I} \mathscr{L}_\alpha = \mathbb{P}\Big(\bigcap_{\alpha\in I} L_\alpha\Big) \cap \mathrm{Gr}(k,d) = \mathbb{P}\Big(\bigcap_{\alpha\in I_r} L_\alpha\Big) \cap \mathrm{Gr}(k,d) = \bigcap_{\alpha\in I_r} \mathscr{L}_\alpha,$$

which proves the claim. □

Corollary 8.5 *The family of linear arrangements in* $\mathrm{Gr}(k,d)$ *is closed under finite unions and arbitrary intersections.*

Proof The statement about finite unions and finite intersections is an immediate consequence of the definition of linear arrangement. The claim for arbitrary intersections follows directly from Lemma 8.4. □

Corollary 8.6 *Let $\mathscr{L}$ be a linear arrangement in* $\mathrm{Gr}(k,d)$ *and let* $B \in \mathrm{GL}(d)$. *If* $B(\mathscr{L}) \subset \mathscr{L}$ *then* $B(\mathscr{L}) = \mathscr{L}$.

Proof Consider the non-increasing sequence $B^n(\mathscr{L})$, $n \geq 1$. By Lemma 8.4 there exists $p \geq 1$ such that $B^n(\mathscr{L}) = B^p(\mathscr{L})$ for every $n \geq p$. In particular, $B^{p+1}(\mathscr{L}) = B^p(\mathscr{L})$ and, applying B^{-p} to both sides, this gives $B(\mathscr{L}) = \mathscr{L}$. □

8.3.2 Linear arrangements and the twisting property

A linear arrangement is *proper* if it is neither empty nor the whole $\mathrm{Gr}(k,d)$.

Lemma 8.7 *A monoid $\mathscr{B}$ is twisting if and only if, for any $1 \leq k \leq d-1$, there is no proper linear arrangement $\mathscr{L}$ in* $\mathrm{Gr}(k,d)$ *with $B(\mathscr{L}) = \mathscr{L}$ for all $B \in \mathscr{B}$.*

Proof Suppose that there exists a proper linear arrangement $\mathscr{L} \subset \mathrm{Gr}(k,d)$ invariant under every $B \in \mathscr{B}$. Fix $F \in \mathscr{L}$ and consider $G_1, \ldots, G_m \in \mathrm{Gr}(d-k,d)$ such that $\mathscr{L}$ is contained in the union of the hyperplane sections $\mathscr{S}_1, \ldots, \mathscr{S}_m$ dual to $G_1, \ldots, G_m$. Then, for every $B \in \mathscr{B}$, there exists j such that $B(F) \in \mathscr{S}_j$ or, equivalently, $F \cap G_j \neq \{0\}$. This shows that $\mathscr{B}$ is not twisting.

Conversely, suppose that $\mathscr{B}$ is not twisting. Then, there exist F, $G_1, \ldots, G_m$ such that, for every $B \in \mathscr{B}$, there exists $j \in \{1,\ldots,m\}$ such that $B(F)$ belongs to the hyperplane section $\mathscr{S}_j$ dual to G_j. Consider

$$\mathscr{L} = \bigcap_{B \in \mathscr{B}} B^{-1}\big(\mathscr{S}_1 \cup \cdots \cup \mathscr{S}_m\big).$$

Then $\mathscr{L}$ is a linear arrangement, by Corollary 8.5, and it is proper, since $F \in \mathscr{L}$. Moreover, $B^{-1}(\mathscr{L}) \supset \mathscr{L}$ for every $B \in \mathscr{B}$. Using Corollary 8.6 we conclude that $B(\mathscr{L}) = \mathscr{L}$ for every $B \in \mathscr{B}$. This finishes the proof. □

Corollary 8.8 *The inverse monoid $\mathscr{B}^{-1} = \{B^{-1} : B \in \mathscr{B}\}$ is pinching or twisting if and only if $\mathscr{B}$ is pinching or twisting, respectively.*

Proof By Exercise 8.2, any B and B^{-1} have the same singular values. Thus, $\mathscr{B}^{-1}$ is pinching if and only if $\mathscr{B}$ is pinching. The claim about twisting is a direct consequence of Lemma 8.7, because $B(\mathscr{L}) = \mathscr{L}$ if and only if $B^{-1}(\mathscr{L}) = \mathscr{L}$. □

The proof of the following corollary mimics that of Lemma 6.9.

Corollary 8.9 *If $\mathscr{B}$ is twisting then any F_k-stationary measure gives zero weight to every proper linear arrangement in* $\mathrm{Gr}(k,d)$.

Proof Suppose $\eta(\mathscr{S}) > 0$ for some linear section $\mathscr{S} \subset \mathrm{Gr}(k,d)$. By definition, $\mathscr{S} = \mathrm{Gr}(k,d) \cap \mathbb{P}S$ for some geometric subspace $S \subset \Lambda^k E$. Let $d_0 \geq 1$ be the smallest dimension of a geometric subspace S such that $\eta(\mathrm{Gr}(k,d) \cap \mathbb{P}S) > 0$ and let $c > 0$ be the supremum of $\eta(\mathrm{Gr}(k,d) \cap \mathbb{P}S)$ over all subspaces S of dimension d_0.

Let $\mathscr{V}$ be the family of all $\mathscr{S} = \mathrm{Gr}(k,d) \cap \mathbb{P}S$ such that S is a geometric subspace with $\dim S = d_0$ and $\eta(\mathscr{S}) = c$. We claim that $\mathscr{V}$ is non-empty. Indeed, consider any sequence $\mathscr{S}_n = \mathrm{Gr}(k,d) \cap \mathbb{P}S_n$ where S_n is a geometric subspace with $\dim S_n = d_0$ and $\eta(\mathscr{S}_n) \to c$. Up to restricting to a subsequence, we may suppose (Exercise 8.8) that $(S_n)_n$ converges to some geometric subspace S of dimension d_0. Since $\mathscr{S} = \mathrm{Gr}(k,d) \cap \mathbb{P}S$ is a closed subset of the Grassmannian, it follows that $\eta(\mathscr{S}) = c$. This proves our claim.

We are also going to show that $\mathscr{V}$ is finite. Indeed, if $\mathscr{S}_i = \mathrm{Gr}(k,d) \cap \mathbb{P}S_i$, $i = 1,2$ are two distinct elements of $\mathscr{V}$ then $\mathscr{S}_1 \neq \mathscr{S}_2$ then $\dim(S_1 \cap S_2) < d_0$ and so $\mathscr{S}_1 \cap \mathscr{S}_2 = \mathrm{Gr}(k,d) \cap \mathbb{P}(S_1 \cap S_2)$ has zero η-measure. Therefore, $\#\mathscr{V} \leq 1/c < \infty$.

Now, consider any $\mathscr{S} \in \mathscr{V}$. Since η is F_k-stationary,

$$c = \eta(\mathscr{S}) = \int \eta(B^{-1}(\mathscr{S}))\, d\nu(B)$$

and so $\eta(B^{-1}(\mathscr{S})) = c$ for ν-almost every B. It follows that $\eta(B^{-1}(\mathscr{S})) = c$ for every $B \in \operatorname{supp} \nu$. This proves that $B^{-1}(\mathscr{S}) \in \mathscr{V}$ for all $B \in \mathscr{B}$ and $\mathscr{S} \in \mathscr{V}$. Consequently, $\mathscr{L}_0 = \bigcup_{\mathscr{S} \in \mathscr{V}} \mathscr{S}$ is a proper linear arrangement in $\mathrm{Gr}(k,d)$, and $\mathscr{L}_0$ is invariant under every $B \in \mathscr{B}$. By Lemma 8.7, this contradicts the assumption that $\mathscr{B}$ is twisting. This contradiction proves that $\eta(\mathscr{S}) = 0$ for every linear section $\mathscr{S}$, and so $\eta(\mathscr{L}) = 0$ for every linear arrangement $\mathscr{L} \subset \mathrm{Gr}(k,d)$. □

8.3.3 Control of eccentricity

Let $1 \leq k \leq d-1$ be fixed. Recall that $E_k^u(B) \in \mathrm{Gr}(k,d)$ is the subspace of dimension k most expanded by B and $E_k^s(B) \in \mathrm{Gr}(d-k,d)$ is the subspace of dimension $d-k$ most contracted by B. They are defined when $\sigma_k(B) > \sigma_{k+1}(B)$.

Lemma 8.10 *Let $(B_n)_n$ be a sequence in* $\mathrm{GL}(d)$ *with $\sigma_k(B_n)/\sigma_{k+1}(B_n) \to \infty$ and $B_n(E_k^u(B_n)) \to E_k^u$ and $E_k^s(B_n) \to E_k^s$ as $n \to \infty$. Let K be any compact subset of* $\mathrm{Gr}(k,d)$ *which does not intersect the hyperplane section dual to E_k^s. Then $B_n(K) \to E_k^u$ as $n \to \infty$.*

Proof Suppose that the conclusion is not true. Then, restricting to a subsequence if necessary, there exist $F_n \in K$ such that $B_n(F_n)$ converges to some $F' \neq E^u_k$. Take $h_n \in F_n$ with $\|h_n\| = 1$ such that

$$\frac{B_n(h_n)}{\|B_n(h_n)\|} \to h \notin E^u_k.$$

By Exercise 8.3, there exists $c_1 > 0$ depending only on the distance from K to the hyperplane section dual to E^s_k such that

$$\|B_n(h_n)\| \geq c_1 \sigma_k(B_n) \quad \text{for every } n. \tag{8.6}$$

Since (Exercise 8.2) $B_n(E^u_k(B_n)) = E^s_{d-k}(B_n^{-1})$, there exists $c_2 > 0$ depending only on the distance from h to E^u_k such that

$$1 = \|h_n\| = \|B_n^{-1}(B_n(h_n))\| \geq c_2 \sigma_{d-k}(B_n^{-1}) \|B_n(h_n)\|.$$

for all large n. Consequently, since $\sigma_{d-k}(B_n^{-1}) = \sigma_{k+1}(B_n)^{-1}$ (Exercise 8.2),

$$1 \geq c_1 c_2 \frac{\sigma_k(B_n)}{\sigma_{k+1}(B_n)}$$

for all n large, which contradicts the hypothesis. □

Lemma 8.11 *Let $(B_n)_n$ be a sequence in $\mathrm{GL}(d)$. Suppose that there is $F \in \mathrm{Gr}(k,d)$ such that the set $\{F' \in \mathrm{Gr}(k,d) : B_n(F') \to F\}$ is not contained in a hyperplane section. Then $\sigma_k(B_n)/\sigma_{k+1}(B_n) \to \infty$ and $F = \lim_n B_n(E^u_k(B_n))$.*

Proof Suppose that $\sigma_k(B_n)/\sigma_{k+1}(B_n)$ is bounded along some subsequence. Passing to a subsequence, and replacing B_n by $Y_n B_n Z_n$, with convenient Y_n, $Z_n \in \mathrm{GL}(d)$ such that $\|Y_n^{\pm 1}\|$ and $\|Z_n^{\pm 1}\|$ are bounded, we may assume that there exist $l < k < m$ such that

- $\sigma_j(B_n)/\sigma_k(B_n) \to \infty$ for $j \in \{1, \dots, l\}$,
- $\sigma_j(B_n) = \sigma_k(B_n)$ for $j \in \{l+1, \dots, m\}$,
- $\sigma_j(B_n)/\sigma_k(B_n) \to 0$ for $j \in \{m+1, \dots, d\}$,

and there exists an orthonormal basis $\{e_1, \dots, e_d\}$, independent of n, such that

- $B_n(e_i) = \sigma_i(B_n) e_i$ for every i and n.

Let E^u, E^c, E^s be the spans of $\{e_1, \dots, e_l\}$, $\{e_{l+1}, \dots, e_m\}$, $\{e_{m+1}, \dots, e_d\}$, respectively. We claim that the set of $F' \in \mathrm{Gr}(k,d)$ such that $B_n(F') \to F$ is contained in the hyperplane section dual to some $G \in \mathrm{Gr}(k,d)$. We split the argument into three cases.

If F is not contained in $E^u \oplus E^c$, take G to be any subspace of dimension $d-k$ containing E^s. Observe that $F' \cap E^s \neq \{0\}$, necessarily, and so $F' \cap G \neq \{0\}$. That proves the claim in this case.

From now on, suppose $F \subset E^u \oplus E^c$. If F does not contain E^u, take G to be any subspace of dimension $d-k$ contained in $E^c \oplus E^s$. Observe that F' cannot be transverse to $E^c \oplus E^s$, and so $\dim F' \cap (E^c \oplus E^s) > k-l$. Then

$$\dim F' \cap (E^c \oplus E^s) + \dim G > k - l + d - k = \dim E^c \oplus E^s$$

and that implies that $F' \cap G \neq \{0\}$.

Finally, suppose $E^u \subset F \subset E^c \oplus E^u$. Then $F = E^u \oplus F^c$ for some subspace $F^c \subset E^c$ with dimension $k-l$. Take G to be any subspace of dimension $d-k$ that contains E^s and intersects F^c. Observe that F' must be transverse to $E^c \oplus E^s$ and $F' \cap (E^c \oplus E^s)$ must be the graph of a linear map $F^c \to E^s$. Hence, $F' \cap G \neq \{0\}$. The proof of our claim is complete.

Thus, we have shown that if $\{F' \in \mathrm{Gr}(k,d) : B_n(F') \to F\}$ is not contained in a hyperplane section, for some $F \in \mathrm{Gr}(k,d)$, then $\sigma_k(B_n)/\sigma_{k+1}(B_n) \to \infty$. Moreover, passing to a subsequence if necessary, we may assume that $E^s_k(B_n)$ converges to some E^s_k. Since $\{F' \in \mathrm{Gr}(k,d) : B_n(F') \to F\}$ is not contained in the hyperplane section dual to E^s_k, Lemma 8.10 gives that $F = \lim B_n(E^u_k(B_n))$. That completes the argument. □

The statement and proof of the next result are akin to those of Lemma 6.14.

Lemma 8.12 *Let $(B_n)_n$ be a sequence in* $\mathrm{GL}(d)$ *and β be any probability measure on* $\mathrm{Gr}(k,d)$.

1. *Assume that β gives zero weight to every hyperplane section and assume $\sigma_k(B_n)/\sigma_{k+1}(B_n) \to \infty$ and $B_n(E^u_k(B_n)) \to E^u_k$. Then $(B_n)_*\beta$ converges in the weak* topology to a Dirac mass on E^u_k.*
2. *Assume that β is not supported in a hyperplane section and assume $(B_n)_*\beta$ converges in the weak* topology to a Dirac mass on some $E^u_k \in \mathrm{Gr}(k,d)$. Then $\sigma_k(B_n)/\sigma_{k+1}(B_n) \to \infty$ and $B_n(E^u_k(B_n)) \to E^u_k$.*

Proof Let us first prove (1). Up to restricting to some subsequence of an arbitrary subsequence, we may assume that $E^s_k(B_n)$ also converges to some E^s_k. Take a compact set K disjoint from the hyperplane section dual to E^s_k and such that $\beta(K) > 1-\varepsilon$. Then $(B_n)_*\beta(B_n(K)) > 1-\varepsilon$ and, by Lemma 8.10, $B_n(K)$ is close to E^u_k for all large n. This shows that $(B_n)_*\beta$ converges to the Dirac measure at E^u_k. Now we prove (2). For each $m \geq 1$, let V_m denote the neighborhood of radius $1/m$ around E^u_k. The hypothesis implies that for each $m \geq 1$ there exists $n_m \geq 1$ such that

$$\beta\left(B_n^{-1}(V_m)\right) > 1 - 2^{-m} \quad \text{for every } n \geq n_m.$$

Define $X = \bigcup_{l=1}^{\infty} \bigcap_{m=l}^{\infty} B_{n_m}^{-1}(V_m)$. Then $\beta(X) = 1$ and so, by the hypothesis on β, the set X is not contained in any hyperplane section. Moreover, for every

$F' \in X$ one has $B_{n_m}(F') \to E_k^u$ as $m \to \infty$. Using Lemma 8.11, we conclude that $\sigma_k(B_n)/\sigma_{k+1}(B_n) \to \infty$ and $B_n(E_k^u(B_n)) \to E_k^u$ restricted to the subsequence $(n_m)_m$. The same argument can be applied starting from any subsequence of $(B_n)_n$. Thus, we have shown that $\sigma_k(B_n)/\sigma_{k+1}(B_n) \to \infty$ and $B_n(E_k^u(B_n)) \to E_k^u$ restricted to some subsequence of every subsequence. This implies the claim, and so the proof of the lemma is complete. □

8.3.4 Convergence of conditional probabilities

We call *homogeneous measure* any probability measure β in $\mathrm{Gr}(k,d)$ such that $\beta(\mathscr{L}) = 0$ for every proper linear arrangement $\mathscr{L}$. Corollary 8.9 states that every F_k-stationary measure is homogeneous, if the associated monoid $\mathscr{B}$ is twisting. Clearly, the support of a homogeneous measure can never be contained in a hyperplane section.

Lemma 8.13 *There exist $\theta > 0$, $m \geq 1$, and matrices P, Q, Q_i, $1 \leq i \leq m$ in $\mathscr{B}$, with the following properties:*

(1) *P admits an invariant decomposition $\mathbb{R}^d = U_P \oplus S_P$ where $\dim U_P = k$ and all eigenvalues of P restricted to U_P are larger, in norm, than all eigenvalues of P restricted to S_P.*
(2) *$Q(U_P) \cap S_P = \{0\}$*
(3) *For every $G \in \mathrm{Gr}(d-k,d)$ there exists $i \in \{1,\dots,m\}$ such that the angle between $Q_i(U_P)$ and G is at least 2θ.*

Proof Since $\mathscr{B}$ is pinching, we may find $B_n \in \mathscr{B}$ such that $\sigma_k(B_n)/\sigma_{k+1}(B_n)$ goes to infinity. Up to restricting to a subsequence, we may suppose that $B_n(E_k^u(B_n))$ and $E_k^s(B_n)$ converge to subspaces E_k^u and E_k^s, respectively. Since $\mathscr{B}$ is twisting, we may find $B_0 \in \mathscr{B}$ such that $B_0(E_k^u) \cap E_k^s = \{0\}$. We are going to take $P = B_0 B_n$ for some large n. Fix $c > b > a > 0$, large enough so that $P(E_k^u(B_n))$ is contained in the cone $C(E_k^u(B_n), a)$ for every large n. Since $\sigma_k(B_n)$ is much larger than $\sigma_{k+1}(B_n)$, the map $P = B_0 B_n$ sends $C(E_k^u(B_n), c)$ to a thin cone around $P(E_k^u(B_n))$. Hence, assuming n is large enough,

$$P\big(C(E_k^u(B_n), c)\big) \subset C(E_k^u(B_n), b).$$

Then (Exercise 4.13), P admits a dominated decomposition as in claim (1).

Since $\mathscr{B}$ is twisting, there exists $Q \in \mathscr{B}$ as in claim (2). For the same reason, given any $G \in \mathrm{Gr}(d-k,d)$ there exists $Q_G \in \mathscr{B}$ such that $Q_G(U_P) \cap G = \{0\}$. Moreover, one may take $Q_{G'} = Q_G$ for every $G' \in \mathrm{Gr}(d-k,d)$ in a neighborhood of G. So, by compactness of $\mathrm{Gr}(d-k,d)$, one can choose matrices Q_1,

$\dots, Q_m$ such that

$$\max_{1\le i\le m} |\sin\measuredangle\big(Q_i(U_P),G\big)| > 0.$$

Using compactness once more, the expression on the left is bounded below by some $2\theta > 0$. □

By definition, $P = P_1 \cdots P_{N(P)}$ for some $P_j \in \operatorname{supp}\nu$ for $1 \le j \le N(P)$ and $N(P) \ge 1$. Fix any such factorization of P and let $r > 0$. Define V_P to be the set of all products

$$P' = P'_1 \cdots P'_{N(P)} \quad \text{with } P'_j \text{ in the } r\text{-neighborhood of } P_j \text{ for } 1 \le j \le N(P).$$

Define $N(Q)$, V_Q and $N(Q_i)$, V_{Q_i}, $1 \le i \le m$ analogously, starting from factorizations of Q and Q_i, $1 \le i \le m$ instead. Take $r > 0$ to be small enough that the V_{Q_i}, $1 \le i \le m$ are pairwise disjoint and the conclusions of Lemma 8.13 remain true, for slightly smaller $\theta > 0$, when P, Q, and Q_i, $1 \le i \le m$ are replaced by any $P' \in V_P$, $Q' \in V_Q$, and $Q'_i \in V_{Q_i}$, $1 \le i \le m$. For each $l \ge 1$, let V_P^l denote the set of all products

$$\tilde{P} = P^{(1)} \cdots P^{(l)}, \quad \text{with } P^{(j)} \in V_P \text{ for all } 1 \le j \le l.$$

Assuming $r > 0$ is small, every $\tilde{P} \in \bigcup_l V_P^l$ admits a dominated decomposition $\mathbb{R}^d = U_{\tilde{P}} \oplus S_{\tilde{P}}$ as in part (1) of Lemma 8.13 (Exercise 4.13). Moreover, $U_{\tilde{P}}$ is uniformly close to U_P, so that parts (2) and (3) remain valid as well.

Let C_P denote the subset of $x \in M$ such that $A(f^{j-1}(x))$ is in the r-neighborhood of P_j for every $j = 1, \dots, N(P)$. Note that C_P is a cylinder of M, since A is locally constant, and C_P has positive μ-measure, because every $P_j \in \operatorname{supp}\nu$. Analogously, starting from the chosen factorizations of Q and Q_i, $1 \le i \le m$, we define cylinders C_Q and C_{Q_i}, $1 \le i \le m$ with positive μ-measure. For each $l \ge 1$, let C_P^l denote the subset of $x \in M$ such that $f^{jN(P)}(x) \in C_P$ for $j = 0, \dots, l-1$. Let $n_{l,i} = lN(P) + N(Q) + lN(P) + N(Q_i)$ and define $Y_{l,i}$ to be the subset of points $x \in M$ such that

$$\begin{aligned} &f^{-n_{l,i}}(x) \in C_P^l \quad \text{and} \quad f^{-n_{l,i}+lN(P)}(x) \in C_Q \quad \text{and} \\ &f^{-n_{l,i}+lN(P)+N(Q)}(x) \in C_P^l \quad \text{and} \quad f^{-n_{l,i}+lN(P)+N(Q)+lN(P)}(x) \in C_{Q_i}. \end{aligned} \tag{8.7}$$

Then $Y_{l,i}$ is an $n_{l,i}$-cylinder of M with $\mu(Y_{l,i}) > 0$, for every l and i. For each l fixed, the cylinders $Y_{l,i}$, $1 \le i \le m$ are pairwise disjoint.

Given $\varepsilon > 0$, we say a probability measure ρ in $\operatorname{Gr}(k,d)$ is *ε-concentrated* if there exists some ε-neighborhood $B_\varepsilon \subset \operatorname{Gr}(k,d)$ such that $\rho(B_\varepsilon) > 1 - \varepsilon$.

Lemma 8.14 *Given $\varepsilon > 0$ and any homogeneous measure β in* $\operatorname{Gr}(k,d)$*, there is $l_0 \ge 1$ and, given any $B_0 \in \mathscr{B}$, there is $i = i(B_0) \in \{1, \dots, m\}$ such that $B_*\beta$*

is ε*-concentrated, for every* $B = B_0 Q_i' P' Q' P''$ *with* $Q_i' \in V_{Q_i}$, $P' \in V_P^l$, $Q' \in V_Q$, $P'' \in V_P^l$, *and* $l > l_0$.

Proof Since β is homogeneous, it follows from (1) in Lemma 8.13 that, if l is large, most of the mass of $P''_*\beta$ is concentrated near the sum $U_{P''} \in \mathrm{Gr}(k,d)$ of the eigenspaces of P'' associated with the k eigenvalues with largest norms. Then most of the mass of $(Q'P'')_*\beta$ is concentrated near $Q'(U_{P''})$. By (2) in Lemma 8.13, the latter is transverse to $S_{P'}$. So, if l is large then most of the mass of $(P'Q'P'')_*\beta$ is concentrated near the sum $U_{P'} \in \mathrm{Gr}(k,d)$ of the eigenspaces associated with the k eigenvalues of P' with largest norms. Choose $\theta > 0$ as in Lemma 8.13 and let $G_{B_0} \in \mathrm{Gr}(d-k,d)$ be as in Exercise 8.6. The action of all $B_0 \in \mathscr{B}$ is equicontinuous restricted to the set of k-dimensional subspaces whose angle to G_{B_0} is at least θ. Let $i = i(B_0)$ be as in (3) of Lemma 8.13, for $G = G_{B_0}$. Then most of the mass of $(Q_i'P'QP'')_*\beta$ is concentrated near $Q_i(U_{P'})$, and so most of the mass of $B_*\beta$ is concentrated near $B_0 Q_i'(U_{P'})$, if l is large enough. Moreover, the equicontinuity property allows us to take the condition on l uniform on B_0. This proves the lemma. □

In what follows, take $l_0 = l_0(\varepsilon, \beta) \geq 1$ to be as in Lemma 8.14.

Lemma 8.15 *There is* $c_0 > 0$ *and measurable sets* $Z_l \subset M$ *with* $\mu(X_l) > c_0$ *for every* $l \geq 1$ *such that, for every* $\varepsilon > 0$, *every homogeneous measure* β *in* $\mathrm{Gr}(k,d)$, *and every* $x \in Z_l$ *with* $l > l_0$, *there exists* $n > l$ *such that* $A^n(f^{-n}(x))_*\beta$ *is* ε*-concentrated.*

Proof For each $l \geq 1$ and $1 \leq i \leq m$, let $n_{l,i}$ and $Y_{l,i}$ be as defined in (8.7). Define $n_l = \max_i n_{l,i}$ By ergodicity of (f^{-n_l}, μ), for μ-almost every $x \in M$ there exist infinitely many values of $k \geq 1$ such that $f^{-kn_l}(x) \in Y_{l,1} \cup \cdots \cup Y_{l,m}$. Choose $k = k(x)$ minimum with this property and then let $n = kn_l + n_{l,i}$ and

$$B_0 = A^{kn_l}(f^{-kn_l}(x)) \quad \text{and} \quad B = A^n(f^{-n}(x)).$$

The definitions give that $B = B_0 Q_i' P'' Q' P'$ with $P', P'' \in V_P^l$, $Q' \in V_Q$, and $Q_i' \in V_{Q_i}$ for some $i \in \{1, \ldots, m\}$. Moreover (Exercise 8.7),

$$\mu(Z_l) \geq \min_{1 \leq j \leq m} \frac{\mu(Y_{l,j})}{\mu(Y_{l,1}) + \cdots + \mu(Y_{l.m})} = \min_{1 \leq j \leq m} \frac{\mu(C_{Q_j})}{\mu(C_{Q_1}) + \cdots + \mu(C_{Q_m})}$$

where Z_l is the set of points $x \in M$ for which this $i = i(B_0)$. Let $c_0 > 0$ denote the expression on the right-hand side. To conclude, apply Lemma 8.14. □

Proof of Proposition 8.2 Let η be any stationary measure for F_k: stationary measures do exist, by Proposition 5.6. Let m be the lift of η to $M \times \mathrm{Gr}(k,d)$.

By Corollaries 5.23 and 5.24, m is an F_k-invariant u-state and its disintegration is given by

$$m_x = \lim_n A^n(f^{-n}(x))_* \eta \quad \text{for } \mu\text{-almost every } x. \tag{8.8}$$

Let X_l, $l \geq 1$ be as in Lemma 8.15 and Z be the set of points x that belong to X_l for infinitely many values of l. Then $\mu(Z) \geq c_0$ and $A^n(f^{-n}(x))_*\eta$ accumulates at a Dirac mass for all $x \in Z$. So, m_x is a Dirac mass for every $x \in Z$. By the ergodicity of (f, μ), it follows that m_x is a Dirac mass for μ-almost every x. Denote the support of this Dirac mass by $\hat{\xi}^u(x)$. Since m is a u-state, we can factorize $\hat{\xi}^u = \xi^u \circ \pi_-$ for some measurable $\xi^u : M^- \to \mathrm{Gr}(k,d)$. The fact that m is F_k-invariant means that $\hat{\xi}^u(f(x)) = A(x)(\hat{\xi}^u(x))$ for μ-almost every x. That proves claim (1). By Corollary 8.9, the support of η cannot be contained in any hyperplane section. It follows from part (2) of Lemma 8.12 that

$$\frac{\sigma_k(A^n(f^{-n}(x)))}{\sigma_{k+1}(A^n(f^{-n}(x)))} \to \infty \quad \text{and} \quad A^n(f^{-n}(x))E_k^u(A^n(f^{-n}(x))) \to \hat{\xi}^u(x),$$

for μ-almost every x. In view of Exercise 8.2, this is the same as

$$\frac{\sigma_{d-k}(A^{-n}(x))}{\sigma_{d-k+1}(A^{-n}(x))} \to \infty \quad \text{and} \quad E_{d-k}^s(A^{-n}(x)) \to \hat{\xi}^u(x),$$

which is precisely claim (2) in the proposition. Moreover, claim (3) follows directly from the observation that m projects to η in $\mathrm{Gr}(k,d)$ and the support of η is not contained in any hyperplane section. This finishes the proof of Proposition 8.2. □

To deduce Proposition 8.3, consider the inverse $F^{-1} : M \times \mathbb{R}^d \to M \times \mathbb{R}^d$, which is the linear cocycle defined over $f^{-1} : M \to M$ by

$$A^{-1} : M \to \mathrm{GL}(d), \quad A^{-1}(x) = A(f^{-1}(x))^{-1}.$$

The associated monoid is $\mathscr{B}^{-1} = \{B^{-1} : B \in \mathscr{B}\}$ and, by Corollary 8.8, this monoid is pinching and twisting. Strictly speaking, F^{-1} is not locally constant, because $A^{-1}(x)$ depends on the first (not the zeroth) coordinate of $x \in M$. However, the cocycle $F' : M \times \mathbb{R}^d \to M \times \mathbb{R}^d$ defined over $f^{-1} : M \to M$ by

$$A' : M \to \mathrm{GL}(d), \quad A'(x) = A(x)^{-1}$$

is locally constant, and it is conjugate to F^{-1} by the map $(x,v) \mapsto (f^{-1}(x), v)$ on $M \times \mathbb{R}^d$. In particular, the two cocycles have the same associated monoid $\mathscr{B}^{-1}$. Applying Proposition 8.2 to the cocycle F', with k replaced by $d-k$, and then translating the conclusions to F^{-1} through the conjugacy, we obtain the claims in Proposition 8.3.

Now the proof of Theorem 8.1 is complete.

8.4 Notes

Simplicity criteria were first proven in the special case of products of random matrices. Guivarc'h and Raugi [60] obtained sufficient conditions (strong irreducibility together with the contraction property) for the extremal Lyapunov exponents $\lambda_\pm$ to be simple. Using exterior powers as in Section 4.3.2, this yields sufficient conditions for simplicity of all Lyapunov exponents.

A more direct criterion for simplicity of the whole Lyapunov spectrum of a product of random matrices was given by Gol'dsheid and Margulis [58], based on considering the action of the cocycle on the Grassmannian manifolds. The contraction property in [60] is replaced by a more explicit condition on the Zariski closure of the group generated by the support of the probability measure p.

Bonatti, Viana [38] extended the criterion in [60] to Hölder-continuous linear cocycles with invariant holonomies and Avila and Viana [14] obtained a similar extension for [58]. In either case, the base dynamics is assumed to be a hyperbolic homeomorphism endowed with an invariant probability measure with local product structure.

A main application was the proof of the Zorich–Kontsevich conjecture on the Lyapunov spectrum of the Teichmüller flow on strata of Abelian differentials, by Avila and Viana [15]. See also the proofs, by Cambrainha [42], that typical symplectic cocycles have only non-zero Lyapunov exponents, and by Herrera [63], that certain multidimensional continued fraction algorithms have simple Lyapunov spectra.

The presentation in this chapter was adapted from Avila and Viana [14, 15].

8.5 Exercises

Exercise 8.1 Let $m \geq 2$ and $\mathscr{S}$ be the set of all $(A_1,\dots,A_m) \in \mathrm{GL}(d)^m$ such that the monoid generated by the set $\{A_1,\dots,A_m\}$ is pinching and twisting. Prove that $\mathscr{S}$ is open and has full Lebesgue measure in $\mathrm{GL}(d)^m$.

Exercise 8.2 Prove that, for every $B \in \mathrm{GL}(d)$ and $1 \leq k \leq d$,

$$\sigma_k(B^{-1}) = \sigma_{d-k+1}(B)^{-1} \quad \text{and} \quad \sigma_k(B^t) = \sigma_k(B).$$

Moreover, assuming that $\sigma_k(B) > \sigma_{k+1}(B)$, show:

(1) $E^s_k(B)$ and $E^u_k(B)$ are orthogonal complements to each other;
(2) $B(E^s_k(B))$ and $B(E^u_k(B))$ are orthogonal complements to each other;
(3) $E^s_k(B^{-1}) = B(E^u_{d-k}(B))$ and $E^u_k(B^{-1}) = B(E^s_{d-k}(B))$;

(4) $E^s_k(B^t) = B(E^s_k(B))$ and $E^u_k(B^t) = B(E^u_k(B))$.

Exercise 8.3 Let $B \in \mathrm{GL}(d)$ and $1 \le k \le d-1$ be such that $\sigma_k(B) > \sigma_{k+1}(B)$. Prove that:

(1) $\|B(h)\| \ge \sigma_k(x)\|h^+_k\|$ for every $h \in \mathbb{R}^d$, where h^+ represents the component of h along $E^u_k(B)$;
(2) if $F \in \mathrm{Gr}(k,d)$ is transverse to $E^s_k(x)$ then $\|B(h)\| \ge c\sigma_k(B)\|h\|$ for all $h \in F$, where $c > 0$ does not depend on B, but only on the distance between F and the hyperplane section dual to $E^s_k(x)$.

Exercise 8.4 Let $\mathscr{B}$ be a monoid. Prove that:

(1) if there is $B_1 \in \mathscr{B}$ whose eigenvalues are all distinct in norm, then $\mathscr{B}$ is pinching;
(2) if there exists B_1 as above and there exists $B_2 \in \mathscr{B}$ such that $B_2(V) \cap W = \{0\}$ for any pair of B_1-invariant subspaces with complementary dimensions, then $\mathscr{B}$ is twisting.

Exercise 8.5 Show that the adjoint monoid $\mathscr{B}^t = \{B^t : B \in \mathscr{B}\}$ is pinching or twisting if and only if $\mathscr{B}$ is pinching or twisting, respectively.

Exercise 8.6 For each $B \in \mathrm{GL}(d)$ let G_B be some subspace of dimension $d-k$ spanned by eigenvectors associated with the $d-k$ smallest eigenvalues of B^tB. If $\sigma_k(B) > \sigma_{k+1}(B)$ then the only choice is $G_B = E^s_k(B)$. Let $\theta > 0$ be fixed. Show that for any $\varepsilon > 0$ there exists $\delta > 0$ such that

$$|\sin\measuredangle(F_1,F_2)| < \delta \quad \Rightarrow \quad |\sin\measuredangle(B(F_1),B(F_2))| < \varepsilon$$

for any $B \in \mathrm{GL}(d)$ and F_1, $F_2 \in \mathrm{Gr}(k,d)$ with $|\sin\measuredangle(F_i,G_B)| > \theta$ for $i = 1,2$.

Exercise 8.7 Let $f : (M,\mu) \to (M,\mu)$ be a Bernoulli shift, $M = X^{\mathbb{Z}}$ and $\mu = p^{\mathbb{Z}}$. Let $m,n \ge 1$, and $Y_1, \dots, Y_m$ be positive measure subsets of M defined by imposing conditions on coordinates $x_0, \dots, x_{n-1}$ only. Define $g(x) = f^{kn}(x)$ with $k = k(x) \ge 1$ minimum such that $f^{kn}(x) \in Y_1 \cup \cdots \cup Y_m$. Prove that, given any sequence $(i_k)_k$ with values in $\{1,\dots,m\}$,

$$\mu(Z) \ge \min_{1\le j\le m} \frac{\mu(Y_j)}{\mu(Y_1)+\cdots+\mu(Y_m)}$$

where Z is the set of all points $x \in M$ for which $g(x) \in Y_{i_{k(x)}}$.

Exercise 8.8 Prove that:

(1) the set of decomposable k-vectors is a closed subset of $\Lambda^k(\mathbb{R}^d)$ and its intersection with the unit ball is compact;

(2) the set of geometric hyperplanes is a compact subset of the space of hyperplanes of $\Lambda^k(\mathbb{R}^d)$;
(3) for every fixed l, the set of geometric subspaces is a compact subset of the space of l-dimensional subspaces of $\Lambda^k(\mathbb{R}^d)$.

Exercise 8.9 Prove that if $\mathscr{B}$ is pinching and twisting then each associated Grassmannian cocycle $F_k : M \times \mathrm{Gr}(k,d) \to M \times \mathrm{Gr}(k,d)$, $1 \le k \le d-1$ has a unique invariant u-state; namely, the probability measure m^u defined on $M \times \mathrm{Gr}(k,d)$ by $m^u(E) = \mu\big(\{x \in M : (x, \hat{\xi}^u(x)) \in E\}\big)$. Analogously, each F_k has a unique invariant s-state.

9

Generic cocycles

In Chapters 7 and 8 we came to the conclusion that, for a significant class of linear cocycles, the Lyapunov exponents are most of the time distinct. The results were stated for locally constant cocycles over Bernoulli shifts but, as observed at the end of both chapters, the conclusions extend much beyond: roughly speaking, they remain valid for Hölder-continuous cocycles with invariant holonomies, assuming that the base dynamics is sufficiently "chaotic".

Rather in contrast, in the early 1980s Mañé [88] announced that generic (that is, a residual subset of all) area-preserving C^1 diffeomorphisms on any surface have $\lambda_{\pm} = 0$ at almost every point, or else they are Anosov diffeomorphisms. Actually, as observed in Example 2.10, the second alternative is possible only if the surface is the torus $\mathbb{T}^2$. A complete proof of Mañé's claim was first given by Bochi [30], based on an unpublished draft by Mañé himself. This family of ideas is the subject of the present chapter.

In Section 9.1, we make a few useful observations about semi-continuity of Lyapunov exponents. Then, in Section 9.2, we state and prove a version of the Mañé–Bochi theorem for continuous linear cocycles (Theorem 9.5). This can be extended in several ways: to the original setting of diffeomorphisms (derivative cocycles); to higher dimensions; and to continuous time systems. Some of these are briefly discussed in Section 9.2.4.

This theory is also connected to the question of how Lyapunov exponents vary with the linear cocycle, which will be the central topic of Chapter 10. Indeed, Theorem 9.5 yields (in Corollary 9.8) a complete characterization of the continuity points of the Lyapunov exponents in the realm of continuous linear cocycles. Similar ideas allow us, in Section 9.3, to give examples of discontinuity of the Lyapunov exponents among Hölder-continuous cocycles.

9.1 Semi-continuity

Let $f : M \to M$ be a measurable transformation on a separable complete metric space M and let μ be a Borel probability measure on M invariant under f. Let $F : M \times \mathbb{R}^d \to M \times \mathbb{R}^d$ be the linear cocycle defined over (f, μ) by a bounded measurable function $A : M \to \mathrm{GL}(d)$.

To begin with, we are going to show that the extremal Lyapunov exponents are semi-continuous functions of the corresponding cocycle. We write $\lambda_\pm(A, \mu)$ to mean $\lambda_\pm(F, \mu)$. Similarly, let $\chi_1(A, \mu) \geq \cdots \geq \chi_d(A, \mu)$ denote all the Lyapunov exponents of F, counted with multiplicity.

The first lemma and its corollaries hold for pairs (A, μ) in any of the following situations:

- A bounded and continuous, with the topology of uniform convergence (given by the C^0-norm), and $\mu \in \mathcal{M}(M)$ with the weak* topology;
- A bounded and measurable, with the topology of uniform convergence, and $\mu \in \mathcal{M}(M)$ with the pointwise topology.

Lemma 9.1 *The function $(A, \mu) \mapsto \lambda_+(A, \mu)$ is upper semi-continuous and the function $(A, \mu) \mapsto \lambda_-(A, \mu)$ is lower semi-continuous. More generally, for any $0 \leq k \leq d$, the function $(A, \mu) \mapsto \sum_{i=1}^k \chi_i(A, \mu)$ is upper semi-continuous and the function $(A, \mu) \mapsto \sum_{i=d-k+1}^d \chi_i(A, \mu)$ is lower semi-continuous.*

Proof Observe that $(A, \mu) \mapsto \int \log \|A^n\| \, d\mu$ is continuous for every $n \geq 1$. Indeed,

$$\Big| \int \log \|A^n\| \, d\mu - \int \log \|B^n\| \, d\nu \Big| \leq \Big| \int \log \|A^n\| \, d(\mu - \nu) \Big| + \int \big| \log \|A^n\| - \log \|B^n\| \big| \, d\nu$$

and the first term goes to zero when $\mu \to \nu$, because $\log \|A^n\|$ is bounded, and the second term goes to zero when $\|A - B\| \to 0$, because the difference $\log \|A^n\| - \log \|B^n\|$ converges uniformly to zero. So, the case $k = 1$ of the claim follows directly from the identities

$$\lambda_+(A, \mu) = \inf_{n \geq 1} \frac{1}{n} \int \log \|A^n\| \, d\mu \quad \text{and}$$
$$\lambda_-(A, \mu) = \sup_{n \geq 1} \frac{1}{n} \int \log \|(A^n)^{-1}\|^{-1} \, d\mu$$

in Theorem 3.12. To deduce the general case, consider the cocycle $\Lambda^k F$ induced by F in the exterior k-power $\Lambda^k(\mathbb{R}^d)$. From Proposition 4.17 we get

$$\chi_1(A,\mu)+\cdots+\chi_k(A,\mu)=\lambda_+(\Lambda^k A,\mu)$$
$$\chi_{d-k+1}(A,\mu)+\cdots+\chi_d(A,\mu)=\lambda_-(\Lambda^k A,\mu).$$

Then it suffices to observe that $A\mapsto\Lambda^k A$ is continuous with respect to the topology of uniform convergence (Exercise 9.1). □

A subset of a topological space is called *residual* if it contains a countable intersection of open dense subsets, and it is called *meager* if its complement is residual or, in other words, if it is contained in a countable union of nowhere-dense closed sets. The ambient is called a *Baire space* if every residual subset is dense or, equivalently, if every meager subset is nowhere dense. Complete metric spaces and locally compact topological spaces are Baire spaces.

Corollary 9.2 *The set of discontinuity points (A,μ) for the Lyapunov exponents $\lambda_\pm$ is a meager subset of the domain.*

Proof This is a general feature of semi-continuous functions. We recall the argument. Let Q be any countable dense subset of $\mathbb{R}$ and, for each $q\in Q$, define

$$F_q=\partial\{(A,\mu):\lambda_+(A,\mu)<q\}.$$

The assumption gives that F_q is the boundary of an open set, and so it is closed and nowhere dense. Now let (A,μ) be a point of discontinuity for λ_+. Then there exists $(A_n,\mu_n)\to(A,\mu)$ such that $\lim_n\lambda_+(A_n,\mu_n)<\lambda_+(A,\mu)$. Let q be any element of Q between these two numbers. Then $(A,\mu)\in F_q$. This proves that the set of discontinuity points of λ_+ is contained in $\bigcup_{q\in Q}F_q$. Analogously, one gets that the set of discontinuity points of λ_- is contained in a countable union of closed nowhere-dense subsets. □

Corollary 9.3 *If (A,μ) is such that all the Lyapunov exponents are equal, then (A,μ) is a continuity point for the Lyapunov exponents.*

Proof Denote $\lambda=\lambda_+(A,\mu)=\lambda_-(A,\mu)$. By upper semi-continuity of the largest exponent and lower semi-continuity of the smallest exponent,

$$\lambda_+(B,\nu)<\lambda+\varepsilon \quad\text{and}\quad \lambda_-(B,\nu)>\lambda-\varepsilon$$

for every (B,ν) close to (A,μ). Then $|\chi_j(B,\nu)-\chi_j(A,\mu)|=|\chi_j(B,\nu)-\lambda|<\varepsilon$ for every $j=1,\dots,d$, and so (A,μ) is indeed a continuity point. □

Another rather general class of continuity points for the Lyapunov exponents are the hyperbolic cocycles introduced in Section 2.2.

Lemma 9.4 *The Lyapunov exponents vary continuously with $A \in \mathrm{GL}(2)$ in the open subset of hyperbolic cocycles in $C^0(M, \mathrm{SL}(2))$.*

Proof Let $\mathbb{R}^d = E^s_{A,x} \oplus E^u_{A,x}$ be the hyperbolic decomposition for the cocycle defined by A. Given $x \in M$, define

$$g^s_A(x) = \frac{\|A(x)v_s\|}{\|v_s\|} \quad \text{and} \quad g^u_A(x) = \frac{\|A(x)v_u\|}{\|v_u\|}$$

for any non-zero $v_s \in E^s_x$ and $v_u \in E^u_x$ (the definition does not depend on the choice of these vectors). The Lyapunov exponents are given by

$$\lambda_+(A,\mu) = \int \log g^u_A \, d\mu \quad \text{and} \quad \lambda_-(A,\mu) = \int \log g^s_A \, d\mu. \tag{9.1}$$

By Proposition 2.6, the invariant sub-bundles E^s_A and E^u_A depend continuously on A. Thus, g^s_B and g^u_B are uniformly close to g^s_A and g^u_A, respectively, if B is close to A. Then, by (9.1), the exponents $\lambda_\pm(B,\mu)$ are close to $\lambda_\pm(A,\mu)$. □

9.2 Theorem of Mañé–Bochi

An invariant probability measure of a transformation $f : M \to M$ is *aperiodic* if the set of periodic points of has zero measure. Given an invariant set $\Lambda \subset M$, we say that a cocycle $F : M \times \mathbb{R}^d \to M \times \mathbb{R}^d$ is *hyperbolic over* Λ if the restriction of F to $\Lambda \times \mathbb{R}^d$ is hyperbolic.

Theorem 9.5 *Let $f : M \to M$ be a homeomorphism and μ be an aperiodic ergodic probability measure on some compact metric space M. For any continuous function $A : M \to \mathrm{SL}(2)$, either the cocycle associated with A is hyperbolic over the support of μ or A is approximated in $C^0(M, \mathrm{SL}(2))$ by continuous functions $C : M \to \mathrm{SL}(2)$ such that $\lambda_\pm(C,\mu) = 0$.*

Before proving this theorem, let us list a few applications:

Corollary 9.6 *In the setting of Theorem 9.5, the set of $A \in C^0(M, \mathrm{SL}(2))$ for which either the linear cocycle is hyperbolic over the support of μ or else $\lambda_\pm(A,\mu) = 0$ is a residual subset of $C^0(M, \mathrm{SL}(2))$.*

Proof By Proposition 2.6, the subset H of functions A for which the linear cocycle is hyperbolic is open in $C^0(M, \mathrm{SL}(2))$. By Lemma 9.1, the subset $L(\delta)$ of continuous functions $A : M \to \mathrm{SL}(2)$ for which $\lambda_+(A,\mu) < \delta$ is also open. Fix any sequence $(\delta_n)_n \to 0$. Then the set in the statement coincides with $\bigcap_n H \cup L(\delta_n)$. By Theorem 9.5, the intersection is dense and, hence, so is each of the open sets $H \cup L(\delta_n)$. □

In some cases, the hyperbolicity alternative can be excluded a priori. Then one gets an abundance of vanishing exponents:

Example 9.7 Let $S^1 = \mathbb{R}/\mathbb{Z}$ and $f : S^1 \to S^1$ be a continuous transformation, with degree $\deg(f) \in \mathbb{Z}$. Let $A : S^1 \to \mathrm{SL}(2)$ be of the form $A(x) = A_0 R_{2\pi\alpha(x)}$ where $A_0 \in \mathrm{SL}(d)$ and $\alpha : S^1 \to S^1$ is a continuous function with degree $\deg(\alpha) \in \mathbb{Z}$ and R_θ denotes the rotation of angle θ. Assume that μ is supported on the whole S^1. Assume that $2\deg(\alpha)$ is not a multiple of $\deg(f) - 1$. Then, continuous cocycles with zero Lyapunov exponents form a residual subset of the isotopy class of A for, as we have seen in Example 2.9, no element of the isotopy class can be hyperbolic.

Theorem 9.5 also yields a complete characterization of the continuity points for the Lyapunov exponents among continuous cocycles:

Corollary 9.8 *In the setting of Theorem 9.5, a function $A : M \to \mathrm{SL}(2)$ is a continuity point for the Lyapunov exponents in $C^0(M, \mathrm{SL}(2))$ if and only if the corresponding linear cocycle is hyperbolic over the support of μ or else the Lyapunov exponents $\lambda_\pm(A, \mu)$ vanish.*

Proof By Lemma 9.4, Lyapunov exponents are continuous at every hyperbolic cocycle. By Corollary 9.3, the same is true for every cocycle with vanishing Lyapunov exponents. Theorem 9.5 implies that there are no other continuity points. □

Let us give an outline of the proof of Theorem 9.5. We will fill in the details in the next three sections.

Start from any $A \in C^0(M, \mathrm{SL}(2))$ whose associated linear cocycle is not hyperbolic over $\operatorname{supp}\mu$. Proposition 9.10 below states that if the Lyapunov exponents $\lambda_\pm(A, \mu)$ are non-zero then one can find a nearby *measurable* function $B : M \to \mathrm{SL}(2)$ that preserves the *union* of the Oseledets sub-bundles of A but exchanges the two Oseledets sub-bundles over some positive measure set $Z \subset M$.

Another key point, that will be established in Proposition 9.12, is that Z may be chosen so that the *second* return map $Z \to Z$ is ergodic. According to Proposition 9.13 below, it follows that $\lambda_\pm(B, \mu) = 0$. This is not quite the end yet, because B need not be continuous. However, using Lusin's theorem, there exist continuous functions $C : M \to \mathrm{SL}(2)$ with uniformly bounded norm and coinciding with B outside sets that have arbitrarily small measure. Then the Lyapunov exponents $\lambda_\pm(C, \mu)$ are arbitrarily close to zero.

This proves that every non-hyperbolic cocycle is approximated by continuous cocycles with arbitrarily small Lyapunov exponents. A Baire argument wraps up the proof of the theorem.

9.2.1 Interchanging the Oseledets subspaces

Let us now present the arguments in detail.

Lemma 9.9 *Let $f : M \to M$ be an invertible transformation and μ be an invariant aperiodic probability measure. For any measurable set $Y \subset M$ with $\mu(Y) > 0$ and any $m \geq 1$ there exists a measurable set $Z \subset Y$ such that $\mu(Z) > 0$ and Z, $f(Z)$, ..., $f^{m-1}(Z)$ are pairwise disjoint.*

Proof Since fixed points form a zero measure subset of Y, we may find $Y_1 \subset Y$ such that $\mu(Y_1 \Delta f(Y_1)) > 0$. Then $Z_1 = Y_1 \setminus f(Y_1)$ has positive measure and it is disjoint from its image $f(Z_1)$. Similarly, we may find $Y_2 \subset Z_1$ such that $\mu(Y_2 \Delta f^2(Y_2)) > 0$. Then $Z_2 = Y_2 \setminus f^2(Y_2)$ has positive measure and Z_2, $f(Z_2)$, and $f^2(Z_2)$ are pairwise disjoint. Continuing in this way we find $Z = Z_m$ as in the statement. □

Proposition 9.10 *Let $A \in C^0(M, \mathrm{SL}(2))$ be such that the associated linear cocycle F is not hyperbolic over* $\operatorname{supp} \mu$. *Then for every $\varepsilon > 0$ there exists $m \geq 1$ and $Z \subset M$ with $\mu(Z) > 0$ such that Z, $f(Z)$, ..., $f^{m-1}(Z)$ are pairwise disjoint, and there exists a measurable map $B : M \to \mathrm{SL}(2)$, satisfying*

(a) $\|A(x) - B(x)\| < \varepsilon$ *for every $x \in M$ and $A = B$ outside $\bigcup_{j=0}^{m-1} f^j(Z)$;*

(b) $B(x)E_x^s = E_{f^m(x)}^u$ *and* $B(x)E_u^s = E_{f^m(x)}^u$ *for $x \in Z$, where $\mathbb{R}^d = E_x^s \oplus E_x^u$ is the Oseledets decomposition of the linear cocycle F associated with A.*

Proof There are two parts. First, we explain how to construct a perturbation that maps E^u to E^s. More precisely, we find $Y \subset M$, $p \geq 1$, and $J : M \to \mathrm{SL}(2)$ satisfying (a) and $J^p(x)E_x^u = E_{f^p(x)}^s$ but, possibly, not $J^p(x)E_x^s = E_{f^p(x)}^u$. Next, we explain how to modify the construction and obtain Z, m, and B satisfying all the conditions in the statement.

Mapping E^u to E^s: Fix $\delta > 0$, much smaller than $\varepsilon > 0$. If the set of points x such that $|\sin \measuredangle(E_x^u, E_x^s)| < \delta$ has positive measure, we may take Y to be that set, $p = 1$, and $J(x) = A(x)R(x)$ for $x \in Y$, where $R(x)$ is a small rotation such that $R(x)E_x^s = E_x^u$. From now on, we suppose that $|\sin \measuredangle(E_x^u, E_x^s)| \geq \delta$ for almost every x. For each $p \geq 1$, consider the set

$$\Gamma_p = \{x \in M : \|A^p(x) \mid E_x^u\| < 2\|A^p(x) \mid E_x^s\|\}.$$

The assumption that the cocycle is non-hyperbolic implies that $\mu(\Gamma_p) > 0$ for every p. Indeed, suppose there exists $p \geq 1$ such that

$$\frac{\|A^p(x) \mid E^u_x\|}{\|A^p(x) \mid E^s_x\|} \geq 2 \quad \text{for } \mu\text{-almost every } x \in M.$$

Then (Exercise 2.4), for any $k \geq 1$ and μ-almost every x

$$\|A^{kp}(x)\|^2 \geq \frac{\|A^{kp}(x) \mid E^u_x\|}{\|A^{kp}(x) \mid E^s_x\|} \geq 2^k.$$

Then, by continuity, $\|A^{kp}(x)\| \geq 2^{k/2}$ for every $k \geq 1$ and x in the support of μ. It follows, by Proposition 2.1, that the linear cocycle F^p is hyperbolic over $\operatorname{supp}\mu$. Then (Exercise 2.5), F is hyperbolic over $\operatorname{supp}\mu$. This contradiction proves that $\mu(\Gamma_p) > 0$ for every p, as claimed.

Now fix $\theta > 0$ much smaller than δ and $p \geq 1$ much larger than $|\log\theta|$. By Lemma 9.9, we may find a measurable set $Y \subset \Gamma_p$ with $\mu(Y) > 0$ and such that $Y, f(Y), \ldots, f^{p-1}(Y)$ are pairwise disjoint. For each $x \in Y$ and $0 < j < m-1$, define

$$J(f^j(x)) = A(f^j(x))L_j(x),$$

where $L_j(x)$ is a linear map that fixes $E^s_{f^j(x)}$ and $E^u_{f^j(x)}$, expanding the former and contracting the latter by a definite factor $e^{\pm\theta}$:

$$L_j(x) : E^s_{f^j(x)} \oplus E^u_{f^j(x)} \to E^s_{f^j(x)} \oplus E^u_{f^j(x)}, \quad v^s + v^u \mapsto e^{\theta} v^s + e^{-\theta} v^u.$$

Since the angle between the two sub-bundles is bounded from zero, one has that $\|A(f^j(x)) - J(f^j(x))\| < \varepsilon$ for all $0 < j < p-1$ and $x \in Y$, as long as θ is small enough. Then

$$\begin{aligned}\frac{\|A(f^{p-1}(x))J(f^{p-2}(x))\cdots J(f(x))A(x) \mid E^s_x\|}{\|A(f^{p-1}(x))J(f^{p-2}(x))\cdots J(f(x))A(x) \mid E^u_x\|} &\geq e^{2(p-2)\theta}\frac{\|A^p(x) \mid E^s_x\|}{\|A^p(x) \mid E^u_x\|} \\ &\geq e^{2(p-2)\theta-1} > 2\delta^{-1}\theta^{-2},\end{aligned}$$

as long as p is large enough. Since the angle between the Oseledets subspaces is bounded below by δ, this implies that there exists some line $r \subset \mathbb{R}^2$ such that, denoting $r_p = A(f^{p-1}(x))J(f^{p-2}(x))\cdots J(f(x))A(x)r$,

$$|\sin\measuredangle(r, E^u_x)| < \theta \quad \text{and} \quad |\sin\measuredangle(r_p, E^s_{f^p(x)})| < \theta.$$

Define $J(x) = A(x)R(x)$ and $J(f^{p-1}(x)) = L(x)A(f^{p-1}(x))$ where $R(x)$ and $L(x)$ are small rotations mapping E^u_x to r and r_p to $E^s_{f^p(x)}$, respectively. This preserves $\|A - J\| < \varepsilon$ and also gives $J^p(x)E^u_x = E^s_{f^p(x)}$, as required.

Mapping E^s to E^u: The latter property implies that $J^p(x)E^s_x \neq E^s_{f^p(x)}$. Then (Exercise 3.15),

$$\lim_m |\sin\measuredangle(A^{m-p}(f^p(x))J^p(x)E^s_x, E^u_{f^m(x)})| = 0$$

Moreover, by Poincaré recurrence,

$$\limsup_m |\sin\measuredangle(E^u_{f^m(x)}, E^s_{f^m(x)})| > 0.$$

Thus, we may find $m > p$ and a positive measure set $Z \subset Y$ such that, for every $z \in Z$, the angle between $A^{m-p}(f^p(x))J^p(x)E^s_x$ and $E^u_{f^m(x)}$ is much smaller than the angle between the two Oseledets subspaces at $f^m(x)$. hence, there exists a linear map $S(f^m(x))$ close to the identity that fixes $E^s_{f^m(x)}$ and maps $A^{m-p}(f^p(x))J^p(x)E^s_x$ to $E^u_{f^m(x)}$. By Lemma 9.9, Z, $f(Z)$, ..., $f^{m-1}(Z)$ may be assumed to be pairwise disjoint. Let $B : M \to \mathrm{SL}(2)$ be given by

$$B(x) = \begin{cases} J(x) & \text{if } x \in Z \cup f(Z) \cup \cdots \cup f^{p-1}(Z) \\ S(f(x))A(x) & \text{if } x \in f^{m-1}(Z) \\ A(x) & \text{in all other cases.} \end{cases}$$

It is clear from the construction that Z, m, and B satisfy all the properties in the conclusion of the proposition. □

9.2.2 Coboundary sets

We want to deduce from Proposition 9.10 that the Lyapunov exponents of B vanish. The following observation shows that additional information is needed for this.

Let $A : M \to \mathrm{SL}(2)$ be such that the Lyapunov exponents of the associated cocycle are non-zero, and let $\mathbb{R}^d = E^s_x \oplus E^u_x$ be the corresponding Oseledets decomposition. Fix a measurable set $W \subset M$ and let $H(x) : \mathbb{R}^2 \to \mathbb{R}^2$ be defined as follows: if $x \in W$ then $H(x)$ is an orthonormal map that exchanges E^s_x with E^u_x; otherwise, $H(x) = \mathrm{id}$. Then take $B : M \to \mathrm{SL}(2)$ to be given by $B(x) = H(f(x))A(x)H(x)^{-1}$ for every $x \in M$. By construction,

$$\begin{aligned} B(x)E^s_x = E^u_x \quad &\text{and} \quad B(x)E^u_x = E^s_x \quad \text{for } x \in W\Delta f(W) \\ B(x)E^s_x = E^s_x \quad &\text{and} \quad B(x)E^u_x = E^u_x \quad \text{for } x \notin W\Delta f(W). \end{aligned} \tag{9.2}$$

On the other hand, since the norm of $H^{\pm 1}$ is bounded, the Lyapunov exponents of A and B coincide. Hence, the property of exchanging the Oseledets subspaces of A alone cannot force the Lyapunov exponents of B to vanish.

This problem is handled by the notion of coboundary set, introduced by Knill [75] in this context. A measurable set $Z \subset M$ is a *coboundary* for f if there exists some measurable set $W \subset M$ such that Z coincides with $W \Delta f(W)$ up to a zero measure set.

Given any measurable set $Z \subset M$ with $\mu(Z) > 0$, let ν be the normalized restriction of μ to Z, and let $g : Z \to Z$ be the first return map: $g(x) = f^{r(x)}(x)$ where $r(x) \geq 1$ is the first time the forward orbit of x hits z (actually, g is defined on a full measure subset of Z). By Proposition 4.18, the map g is ergodic for the probability measure ν.

Lemma 9.11 *Let $f : M \to M$ be an invertible transformation and μ be an ergodic probability measure. A set $Z \subset M$ with $\mu(Z) > 0$ is a coboundary for f if and only if the second return map $g^2 : Z \to Z$ is not ergodic for ν.*

Proof Suppose that there exists $W \subset M$ be such that $Z = W \Delta f(W)$ up to a zero measure set. Let $U = W \setminus f(W)$ and $V = f(W) \setminus W$. Given $x \in U$, let $n \geq 1$ be minimum such that $f^n(x) \notin W$. Then $g(x) = f^n(x) \in V$. Given $x \in V$, let $n \geq 1$ be minimum such that $f^n(x) \in W$. Then $g(x) = f^n(x) \in U$. This shows that $g(U) \subset V$ and $g(V) \subset U$. It follows that $\nu(U) = \nu(V) = 1/2$ and $g^2(U) \subset U$. In particular, g^2 is not ergodic.

To prove the converse, suppose there exists $U \subset Z$ such that $0 < \nu(U) < 1$ and $g^2(U) = U$. Then $U \cup g(U)$ is a (g, ν)-invariant set with positive measure. Since g is ergodic, it follows that $\nu(U \cup g(U)) = 1$. In other words, $Z = U \cup g(U)$ up to a zero measure set. Define $W = \{f^j(x) : x \in U \text{ and } 0 \leq i < r(x)\}$. Then $f(W) = \{f^j(x) : x \in U \text{ and } 0 < i \leq r(x)\}$ and so

$$W \Delta f(W) = \{f^j(x) : x \in U \text{ and } i \in \{0, r(x)\}\} = U \cup g(U) = Z$$

up to a zero measure set. Thus, Z is a coboundary set. □

Proposition 9.12 *Assume that μ is aperiodic for f. Then any set with positive measure has some subset with positive measure that is not a coboundary set.*

Proof Since M is a compact metric space, the probability space $(M, \mathscr{B}, \mu)$ is a Lebesgue space (see [114, Section 8.5.2]). Moreover, μ is non-atomic, because it is aperiodic. So, (M, μ) is isomorphic to the interval $[0, 1]$, endowed with the Lebesgue measure. Theorem 9.3 in Friedman [53] asserts that, for any ergodic measure-preserving transformation in a Lebesgue space, the family $\mathscr{W}(f, \mu)$, of measurable sets $Z \subset M$ such that the first return map $f_Z : Z \to Z$ is weak mixing, is dense in the σ-algebra of M, relative to the distance $d(A_1, A_2) = \mu(A_1 \Delta A_2)$. In particular, $\mathscr{W}(f, \mu)$ contains some measurable set Z with $\mu(Z) > 0$.

Now, given any $Y \subset M$ with $\mu(Y) > 0$, let $h = f_Y : Y \to Y$ be the first return map and ν be the normalized restriction of μ to Y. Then ν is aperiodic for h. By the previous paragraph, there is $Z \in \mathscr{W}(h, \nu)$ with $\nu(Z) > 0$. This means that $h_Z : Z \to Z$ is weak mixing and $\mu(Z) > 0$. Since weak mixing is preserved under iteration, the second return map h_Z^2 is also weak mixing and, hence, ergodic. Clearly, h_Z coincides with the first return map $f_Z : Z \to Z$ under the transformation f. Thus, f_Z^2 is ergodic and so Z is not a coboundary set. □

Therefore, we may always choose the set Z in Proposition 9.10 in such a way that it is not a coboundary.

Proposition 9.13 *Let $Z \subset M$, $m \geq 1$, and $B : M \to \mathrm{SL}(2)$ be as in the conclusion of Proposition 9.10 and assume that Z is not a coboundary set. Then $\lambda_{\pm}(B,\mu) = 0$.*

Proof Let $\mathbb{R}^d = E_x^s \oplus E_x^u$ denote the Oseledets decomposition for A. Suppose that $\lambda_{\pm}(B,\mu)$ are different from zero and let $\mathbb{R}^d = E_{B,x}^s \oplus E_{B,x}^u$ be the corresponding Oseledets decomposition. We are going to use the inducing construction in Section 4.4.1. More precisely, we consider the linear cocycle

$$G : Z \times \mathbb{R}^2 \to Z \times \mathbb{R}^2, \quad G(x,v) = (f^{r(x)}(x), B^{r(x)}v)$$

over the first return map $g : Z \to Z$. We also consider the corresponding projective cocycle $\mathbb{P}G : Z \times \mathbb{P}\mathbb{R}^2 \to Z \times \mathbb{P}\mathbb{R}^2$. By construction,

$$B^{r(x)}(E_x^s) = E_x^u \quad \text{and} \quad B^{r(x)}(E_x^u) = E_x^s \quad \nu\text{-almost everywhere.} \tag{9.3}$$

By Proposition 4.18, the Lyapunov exponents $\lambda_{\pm}(G,\nu)$ are different from zero, and the Oseledets decomposition of G is given by $\mathbb{R}^d = E_{B,x}^s \oplus E_{B,x}^u$ restricted to the domain Z. Let m be the probability measure defined on $Z \times \mathbb{P}\mathbb{R}^2$ by

$$m(X) = \frac{1}{2}\nu(\{x \in Z : (x, E_x^s) \in X\}) + \frac{1}{2}\nu(\{x \in Z : (x, E_x^u) \in X\}).$$

In other words, m projects down to μ and its disintegration is given by

$$x \mapsto \frac{1}{2}\delta_{E_x^s} + \frac{1}{2}\delta_{E_x^u}.$$

It follows from (9.3) that m is invariant under $\mathbb{P}G$. Then, by Lemma 5.25 and Remark 5.26, m is a linear combination of the probability measures m_B^s and m_B^u defined on $Z \times \mathbb{P}\mathbb{R}^2$ by

$$m_B^s(X) = \nu\big(\{x \in Z : (x, E_{B,x}^s) \in X\}\big)$$
$$m_B^u(X) = \nu\big(\{x \in Z : (x, E_{B,x}^u) \in X\}\big).$$

Lemma 9.14 *The probability measure m is ergodic for $\mathbb{P}G$.*

Proof Suppose that there is an invariant set $X \subset Z \times \mathbb{P}\mathbb{R}^2$ with $0 < m(X) < 1$. Let Z_0 be the set of $x \in Z$ whose fiber $X \cap (\{x\} \times \mathbb{P}\mathbb{R}^2)$ contains neither E_x^s nor E_x^u. In view of (9.3), Z_0 is a (g,ν)-invariant set and so its ν-measure is either

0 or 1. Since $m(X) > 0$, we must have $\nu(Z_0) = 0$. Similarly, $m(X) < 1$ implies that $\nu(Z_2) = 0$, where Z_2 is the set of $x \in Z$ whose fiber contains both E^s_x and E^u_x. Now let Z_s be the set of $x \in Z$ whose fiber contains E^s_x but not E^u_x, and let Z_u be the set of $x \in Z$ whose fiber contains E^u_x but not E^s_x. The previous observations show that $Z_s \cup Z_u$ has full ν-measure. Moreover, (9.3) implies

$$g(Z_s) = Z_u \quad \text{and} \quad g(Z_u) = Z_s$$

up to zero measure sets. Thus, $\nu(Z_s) = \nu(Z_u) = 1/2$ and $g^2(Z_s) = Z_s$. This implies that g^2 is not ergodic, contradicting Lemma 9.11. This contradiction proves the present lemma. □

It follows that m coincides with either m^s_B and m^u_B. This is a contradiction, because the conditional probabilities of m are supported on exactly two points on each fiber, whereas the conditional probabilities of both m^u_B and m^s_B are Dirac masses on a single point. This contradiction proves that the Lyapunov exponents $\lambda_\pm(B,\mu)$ are zero, as claimed. □

9.2.3 Proof of Theorem 9.5

So far we have shown that any continuous cocycle F which is not hyperbolic over the support of μ can be approximated, in the uniform norm, by *measurable* cocycles with vanishing exponents. To complete the proof of Theorem 9.5 we must replace this by *continuous* cocycles. We are going to use the following semi-continuity result:

Lemma 9.15 *Let $B : M \to \mathrm{SL}(d)$, $d \geq 2$ be such that $\log \|B\| \in L^1(\mu)$. Given $L > 0$ and $\delta > 0$, there exists $\rho > 0$ such that $\lambda_+(J,\mu) < \lambda_+(B,\mu) + \delta$ for any measurable function $J : M \to \mathrm{SL}(d)$ such that $\mu\big(\{x \in M : B(x) \neq J(x)\}\big) < \rho$ and $\|J\|_\infty < L$.*

Proof Given $\delta > 0$, fix $n \geq 1$ such that

$$\frac{1}{n}\int \log \|B^n\|\, d\mu \leq \lambda_+(B,\mu) + \frac{\delta}{2}.$$

Given $J : M \to \mathrm{SL}(d)$ such that $\|J\| < L$, denote $\Delta = \{x \in M : B(x) \neq J(x)\}$ and $\Delta^n = \Delta \cup f(\Delta) \cup \cdots \cup f^{n-1}(\Delta)$. Then

$$\lambda_+(J,\mu) \leq \frac{1}{n}\int \log \|J^n\|\, d\mu = \frac{1}{n}\int_{\Delta^n} \log \|J^n\|\, d\mu + \frac{1}{n}\int_{M\setminus\Delta^n} \log \|B^n\|\, d\mu.$$

The right-hand side is bounded by

$$\frac{1}{n}\mu(\Delta^n)\log L + \lambda_+(B,\mu) + \frac{\delta}{2} \leq \mu(\Delta)\log L + \lambda_+(B,\mu) + \frac{\delta}{2}.$$

It follows that $\lambda_+(J,\mu) \le \lambda_+(B,\mu) + \delta$ as long as $\mu(\Delta) < \rho$ for some $\rho > 0$ sufficiently small. This proves the lemma. □

Let us proceed with the proof of Theorem 9.5. Previously, we have shown that for any $\varepsilon > 0$ there exist measurable functions $B : M \to \mathrm{SL}(2)$ such that $\|A - B\|_\infty < \varepsilon$ and $\lambda_+(B,\mu) = 0$. Fix any $\delta > 0$. Given any $\rho > 0$ there exist continuous functions $J : M \to \mathrm{SL}(2)$ such that $\|J\|_\infty \le \|B\|_\infty$ and $J = B$ outside a set with measure ρ (Exercise 9.4). By Lemma 9.15, this implies that $\lambda_+(J,\mu) < \delta$, as long as ρ is chosen small enough.

Let H be the subset of continuous functions $A : M \to \mathrm{SL}(2)$ such that the associated linear cocycle is hyperbolic over $\operatorname{supp}\mu$ and, for any $\delta > 0$, let $L(\delta)$ be the subset of continuous functions $A : M \to \mathrm{SL}(2)$ such that $\lambda_+(A,\mu) < \delta$. By Propositions 2.6 and 9.1, H and the $L(\delta)$ are all open subsets of $C^0(M,\mathrm{SL}(2))$. The reasoning in the previous paragraph shows that $L(\delta)$ is dense in the closed set $C^0(M,\mathrm{SL}(2)) \setminus H$, for every $\delta > 0$. Since $C^0(M,\mathrm{SL}(2))$ is a complete metric space, it follows from Baire's theorem (Exercise 9.3) that $\bigcap_{k=1}^\infty L(1/k)$ is dense in $C^0(M,\mathrm{SL}(2)) \setminus H$. In other words, every non-hyperbolic continuous cocycle is approximated in $C^0(M,\mathrm{SL}(2))$ by another continuous cocycle, whose Lyapunov exponents are zero. This finishes the proof of Theorem 9.5.

9.2.4 Derivative cocycles and higher dimensions

Recall that a C^1 diffeomorphism $f : M \to M$ is *Anosov* if the derivative cocycle $F = Df$ is hyperbolic. We denote by $\mathrm{Diff}^1_m(M)$ the space of volume-preserving diffeomorphisms on a compact Riemannian manifold, endowed with the C^1-topology: two diffeomorphisms are C^1-close if they are uniformly close and their derivatives are also uniformly close. Theorem 9.5 is still true, albeit much harder, for derivative cocycles of area-preserving diffeomorphisms:

Theorem 9.16 (Mañé [88], Bochi [30]) *Let M be a compact Riemannian manifold of dimension 2. For every $f \in \mathrm{Diff}^1_m(M)$, either f is an Anosov diffeomorphism or f is approximated in $\mathrm{Diff}^1_m(M)$ by diffeomorphisms whose Lyapunov exponents vanish almost everywhere.*

Corollary 9.17 *Let M be a compact Riemannian manifold of dimension 2. If $f \in \mathrm{Diff}^1_m(M)$ is a continuity point for the Lyapunov exponents then either f is an Anosov diffeomorphism or the Lyapunov exponents vanish almost everywhere. The continuity points form a residual subset of* $\mathrm{Diff}^1_m(M)$.

The 2-dimensional torus is the only compact surface that carries Anosov diffeomorphisms (see [52, 91] and Example 2.10). In all other cases, Corollary 9.17 gives that a residual subset of all area-preserving diffeomorphisms have zero Lyapunov exponents almost everywhere.

Now we discuss extensions of the previous results to arbitrary dimension. For the statements we need the notion of dominated decomposition. Let M be a compact metric space and $F : M \times \mathbb{R}^d \to M \times \mathbb{R}^d$ be a continuous cocycle over some homeomorphism $f : M \to M$. Let

$$\mathbb{R}^d = V_x^1 \oplus \cdots \oplus V_x^l, \quad x \in \Lambda$$

be an F-invariant decomposition, defined for x in some invariant set $\Lambda \subset M$ and such that every $x \mapsto \dim V_x^i$ is constant on Λ. We call the decomposition *dominated* over Λ if there exist $C > 0$ and $\theta < 1$ such that

$$\frac{\|F^n(x)v_i\|}{\|F^n(x)v_{i-1}\|} \leq C\theta^n \quad \text{for every unit vector } v_i \in V_x^i,\ v_{i-1} \in V_x^{i-1},$$

and for every $x \in X$ and $1 < i \leq l$. In other words, the cocycle is more contractive along each V_x^i than along V_x^{i-1}, by a definite factor. By convention, the trivial decomposition into a single subspace (the case $l = 1$) is dominated.

In the next statement G is any subgroup of $\mathrm{GL}(d)$ that acts *transitively* on the projective space; that is, such that $\{g(r) : g \in G\} = \mathbb{PR}^d$ for any $r \in \mathbb{PR}^d$. That includes many interesting subgroups, such as the special linear group $\mathrm{SL}(d)$ and the symplectic group (if d is even). When d is even, it also includes $\mathrm{SL}(d/2, \mathbb{C})$ and $\mathrm{GL}(d/2, \mathbb{C})$, viewed as subgroups of $\mathrm{GL}(d)$.

Theorem 9.18 (Bochim and Viana [34]) *Let $f : M \to M$ be a homeomorphism on a compact metric space M and μ be an aperiodic ergodic probability measure. Then $A : M \to G$ is a continuity point for the Lyapunov exponents among G-valued continuous cocycles if and only if the Oseledets decomposition of the associated linear cocycle over f is dominated (possibly trivially) over the support of μ. The continuity points form a residual subset of the space of all continuous G-valued cocycles.*

The statement actually proven in [34] is a bit stronger: in particular, it includes the non-ergodic case.

Theorem 9.19 (Bochi and Viana [34]) *For any compact manifold M, there exists a residual subset $\mathcal{R}$ of the space $\mathrm{Diff}_m^1(M)$ of volume-preserving diffeomorphisms such that for every $f \in \mathcal{R}$ the Oseledets decomposition of $F = Df$ is dominated (possibly trivially) on every orbit of f.*

Example 9.20 Bonatti and Viana [38] constructed open sets $\mathcal{U}$ of volume-preserving diffeomorphisms on $M = \mathbb{T}^4$ that are not Anosov and yet admit a *unique* decomposition $TM = E^s \oplus E^u$ that is invariant and dominated for the derivative cocycle. Both subspaces E^s and E^u are 2-dimensional. Tahzibi [110] showed that every C^2 diffeomorphism in $\mathcal{U}$ is ergodic. By Avila [8], the set of C^2 diffeomorphisms is dense in $\mathrm{Diff}^1_m(M)$. By Oxtoby and Ulam [93], the set of ergodic diffeomorphisms is always a countable intersection of open sets. Combining these three facts we get that ergodic diffeomorphisms form a residual subset $\mathcal{R}$ of $\mathcal{U}$. For any $f \in \mathcal{R}$, Theorem 9.19 and Exercise 9.5 give that the Oseledets decomposition extends continuously to a dominated decomposition on the whole M. Then, by uniqueness, the Oseledets decomposition must coincide with $E^s \oplus E^u$ at almost every point. In particular, every $f \in \mathcal{R}$ has exactly one positive exponent and one negative exponent, both with multiplicity 2.

Now take M to be endowed with a symplectic form; that is, a closed non-degenerate differential 2-form ω. Existence of such a form implies that the dimension of M is even, $d = 2k$. Moreover, $\omega^k = \omega \wedge \cdots \wedge \omega$ is a volume form on M. Let m be the corresponding volume measure, normalized so that $m(M) = 1$. A diffeomorphism $f : M \to M$ is *symplectic* if the derivative preserves the symplectic form; that is, if

$$\omega_x(v,w) = \omega_{f(x)}(Df(x)v, Df(x)w) \quad \text{for every } x \in M \text{ and } v,w \in T_xM.$$

Every symplectic diffeomorphism preserves the volume measure m. The space $\mathrm{Diff}^1_\omega(M)$ of symplectic diffeomorphisms is a closed subset of $\mathrm{Diff}^1_m(M)$. In dimension $d = 2$ the two spaces coincide.

Theorem 9.21 (Bochi and Viana [34], Bochi [31]) *There exists a residual subset $\mathcal{R}$ of the space $\mathrm{Diff}^1_\omega(M)$ of symplectic diffeomorphisms such that for every $f \in \mathcal{R}$ the Oseledets decomposition of $F = Df$ is dominated (possibly trivially) on every orbit of f.*

As a consequence, one gets the following dichotomy for a residual subset of symplectic diffeomorphisms that was also announced by Mañé [88]:

(1) either f is an Anosov diffeomorphism, with invariant decomposition $TM = E^u \oplus E^s$ such that $\dim E^u = \dim E^s$ and $Df \mid E^s$ is uniformly contracting and $Df \mid E^u$ is uniformly expanding;
(2) or, for almost every x, there is a decomposition $E^s \oplus E^c \oplus E^u$ dominated on the orbit of x, with $\dim E^s = \dim E^u$ (possibly equal to zero) and $\dim E^c \geq 2$, such that $Df \mid E^s$ is uniformly contracting, $Df \mid E^u$ is uniformly expanding, and the Lyapunov exponents of $Df \mid E^c$ vanish.

For continuous time systems new difficulties arise, especially in the presence of equilibrium points of the flow. In his thesis, Bessa [20] obtained a version of Theorem 9.5 for 2-dimensional continuous time linear cocycles. The extension to arbitrary dimension, corresponding to Theorem 9.18, was obtained by Bessa [22]. As happens for the discrete time systems, the case of derivative cocycles is much harder. A version of Theorem 9.16 for 3-dimensional flows was proven by Bessa [21], in the case when the flow has no equilibrium points. This restriction was removed soon afterwards by Araújo and Bessa [3]. Concerning extensions to higher dimensions, the continuous time version of Theorem 9.19 was settled by Bessa and Rocha [24] and a version of Theorem 9.21 for Hamiltonian flows with two degrees of freedom was obtained by Bessa and Dias [23].

9.3 Hölder examples of discontinuity

Corollary 9.8 shows that continuity of the Lyapunov exponents, although corresponding to residual subsets of the domain (by Corollary 9.2), can only occur at cocycles that are rather special from a dynamical viewpoint. While this is perhaps specific to the C^0-topology in the space of cocycles (see the discussion in Section 10.6), it is interesting to point out that a variation of the proof of Theorem 9.5 also yields examples of discontinuity in the space of Hölder-continuous cocycles over a Bernoulli shift, with arbitrary Hölder constant. That is the purpose of the present section and also motivates some of the ideas we will put forward in Section 10.6.

Let $f : M \to M$ be the shift map on the space $M = X^{\mathbb{Z}}$, with $X = \{1,2\}$, endowed with the metric

$$d(x,y) = 2^{-N(x,y)}, \quad N(x,y) = \sup\{N \geq 0 : x_n = y_n \text{ whenever } |n| < N\}.$$

Let $\mu = p^{\mathbb{Z}}$, where $p = p_1\delta_1 + p_2\delta_2$ with p_1, p_2 positive and distinct.

Let $r \in (0,\infty)$ be fixed. A function $A : M \to \mathrm{GL}(d)$ is *r-Hölder-continuous* if there exists $C > 0$ such that $\|A(x) - A(y)\| \leq Cd(x,y)^r$ for any $x, y \in M$. The space $H^r(M)$ of all r-Hölder-continuous functions $A : M \to \mathrm{GL}(d)$ comes with a natural topology, defined by the *r-Hölder norm*

$$\|A\|_r^H = \sup\Big\{\|A(x)\| : x \in M\Big\} + \sup\Big\{\frac{\|A(x) - A(y)\|}{d(x,y)^r} : x \neq y\Big\}. \tag{9.4}$$

Given $\sigma > 1$, consider the locally constant cocycle associated with the function $A : M \to \mathrm{SL}(2)$ given by

$$A(x) = \begin{pmatrix} \sigma & 0 \\ 0 & \sigma^{-1} \end{pmatrix} \quad \text{if } x_0 = 1$$

and

$$A(x) = \begin{pmatrix} \sigma^{-1} & 0 \\ 0 & \sigma \end{pmatrix} \quad \text{if } x_0 = 2.$$

Note that A is r-Hölder continuous for any $r > 0$. The Lyapunov exponents $\lambda_\pm(A,\mu) = \pm|p_1 - p_2| \log \sigma$ are non-zero.

Theorem 9.22 *For any $r > 0$ such that $2^{2r} < \sigma$ there exist r-Hölder-continuous cocycles $B : M \to \mathrm{SL}(2)$ with vanishing Lyapunov exponents and such that $\|A - B\|_r^H$ is arbitrarily close to zero. Hence, A is a point of discontinuity for the Lyapunov exponents on $H^r(\{1,2\})$.*

The hypothesis on the Hölder constant r is probably not sharp. For reasons to be discussed in Section 10.6, it would be interesting to weaken it to $2^r < \sigma^2$.

Here is an outline of the proof of Theorem 9.22. Note that the original cocycle A preserves the horizontal and vertical line bundles $H_x = \mathbb{R}(1,0)$ and $V_x = \mathbb{R}(0,1)$. Then the Oseledets subspaces must coincide with H_x and V_x almost everywhere. We choose cylinders $Z_n \subset M$ whose first n iterates $f^i(Z_n)$, $0 \le i \le n-1$ are pairwise disjoint. Then we construct cocycles B_n by modifying A on some of these iterates so that

$$B_n^n(x) H_x = V_{f^n(x)} \quad \text{and} \quad B_n^n(x) V_x = H_{f^n(x)} \quad \text{for all } x \in Z_n.$$

We deduce from this property that the Lyapunov exponents of B_n vanish. Moreover, by construction, each B_n is constant on every atom of some finite partition of M into cylinders. In particular, B_n is Hölder continuous for every $r > 0$. From the construction we also get that

$$\|B_n - A\|_r^H \le \mathrm{const}\,\left(2^{2r}/\sigma\right)^{n/2} \tag{9.5}$$

decays to zero as $n \to \infty$. This is how we get the claims in the theorem.

In the remainder of this section we fill the details in this outline, to prove Theorem 9.22. Let $n = 2k+1$ for some $k \ge 1$ and $Z_n = [0; 2, \ldots, 2, 1, \ldots, 1, 1]$ where the symbol 2 appears k times and the symbol 1 appears $k+1$ times. Note that the $f^i(Z_n)$, $0 \le i \le 2k$ are pairwise disjoint. Let

$$\varepsilon_n = \sigma^{-k} \quad \text{and} \quad \delta_n = \arctan \varepsilon_n. \tag{9.6}$$

Define $R_n : M \to \mathrm{SL}(2)$ by

$$R_n(x) = \text{rotation of angle } \delta_n \quad \text{if } x \in f^k(Z_n)$$

$$R_n(x) = \begin{pmatrix} 1 & 0 \\ \varepsilon_n & 1 \end{pmatrix} \quad \text{if } x \in Z_n \cup f^{2k}(Z_n)$$

$$R_n(x) = \mathrm{id} \quad \text{in all other cases.}$$

and then take $B_n = AR_n$.

Lemma 9.23 *$B_n^n(x)H_x = V_{f^n(x)}$ and $B_n^n(x)V_x = H_{f^n(x)}$ for all $x \in Z_n$.*

Proof Note that for any $x \in Z_n$,

$$B_n^k(x)H_x = \mathbb{R}(\varepsilon_n, 1) \quad \text{and} \quad B_n^k(x)V_x = V_{f^k(x)}$$
$$B_n^{k+1}(x)H_x = V_{f^{k+1}(x)} \quad \text{and} \quad B_n^{k+1}(x)V_x = \mathbb{R}(-\varepsilon_n, 1)$$
$$B_n^{2k}(x)H_x = V_{f^{2k}(x)} \quad \text{and} \quad B_n^{2k}(x)V_x = \mathbb{R}(-1, \varepsilon_n).$$

The claim follows by iterating one more time. □

Lemma 9.24 *There is $C > 0$ such that $\|B_n - A\|_r^H \le C\left(2^{2r}/\sigma\right)^k$ for every n.*

Proof Let $L_n = B_n - A$. Clearly, $\sup\|L_n\| \le \sup\|A\|\,\|R_n - \mathrm{id}\|$ and this is bounded by $\sigma\varepsilon_n$. Now let us estimate the second term in the definition (9.4). If x and y are not in the same cylinder $[0;a]$ then $d(x,y) = 1$, and so

$$\frac{\|L_n(x) - L_n(y)\|}{d(x,y)^r} \le 2\sup\|L_n\| \le 2\sigma\varepsilon_n. \tag{9.7}$$

From now on we suppose x and y belong to the same cylinder. Then, since A is constant on cylinders,

$$\frac{\|L_n(x) - L_n(y)\|}{d(x,y)^r} = \frac{\|A(x)(R_n(x) - R_n(y))\|}{d(x,y)^r} \le \sigma\frac{\|R_n(x) - R_n(y)\|}{d(x,y)^r}.$$

If neither x nor y belong to $Z_n \cup f^k(Z_n) \cup f^{2k}(Z_n)$ then $R_n(x) = R_n(y)$ and so the expression on the right vanishes. The same holds if x and y belong to the same $f^i(Z_n)$ with $i \in \{0,k,2k\}$. We are left to consider the case when one of the points belongs to some $f^i(Z_n)$ with $i \in \{0,k,2k\}$ and the other one does not. Then $d(x,y) \ge 2^{-2k}$ and so, using once more that $\|R_n - \mathrm{id}\| \le \varepsilon_n$ at every point, we have

$$\frac{\|L_n(x) - L_n(y)\|}{d(x,y)^r} \le \sigma\frac{\|R_n(x) - R_n(y)\|}{d(x,y)^r} \le 2\sigma\varepsilon_n 2^{2kr}.$$

Combining this with (9.7), we conclude that

$$\|L_n\|_r \le \sigma\varepsilon_n + 2\sigma\varepsilon_n 2^{2kr} \le 3\sigma\left(2^{2r}/\sigma\right)^k.$$

Now it suffices to take $C = 3\sigma$ to complete the proof of the lemma. □

Now we want to prove that $\lambda_\pm(B_n) = 0$ for every n. Let μ_n be the normalized restriction of μ to Z_n and $f_n : Z_n \to Z_n$ be the first return map (defined on a full measure subset). We have the following explicit description:

$$Z_n = \bigsqcup_{b\in\mathscr{B}} [0;w,b,w] \quad \text{(up to a zero measure subset)}$$

where $w = (2,\dots,2,1,\dots,1,1)$ and the union is over the set Ω of all finite words $b = (b_1,\dots,b_s)$ not having w as a subword; moreover,

$$f_n \mid [0;w,b,w] = f^{n+s} \mid [0;w,b,w] \quad \text{for each } b \in \Omega.$$

Thus, (f_n, μ_n) is a Bernoulli shift with an infinite alphabet Ω and probability vector given by $p_b = \mu_n([0;w,b,w])$. Let $\hat{B}_n : Z_n \to \mathrm{SL}(2)$ be the cocycle induced by B over f_n; that is,

$$\hat{B}_n \mid [0;w,b,w] = B_n^{n+s} \mid [0;w,b,w] \quad \text{for each } b \in \Omega.$$

By Proposition 4.18, the Lyapunov spectrum of the induced cocycle is obtained multiplying the Lyapunov spectrum of the original cocycle by the average return time. In our setting this means that

$$\lambda_\pm(\hat{B}_n) = \frac{1}{\mu(Z_n)} \lambda_\pm(B_n).$$

Therefore, it suffices to prove that $\lambda_\pm(\hat{B}_n) = 0$ for every n.

Suppose that the Lyapunov exponents of $\hat{B}_n$ are different from zero and let $\mathbb{R}^d = E_x^u \oplus E_x^s$ be the Oseledets decomposition (defined almost everywhere in Z_n). The key observation is that, as a consequence of Lemma 9.23, the cocycle $\hat{B}_n$ permutes the vertical and horizontal sub-bundles:

$$\hat{B}_n(x) H_x = V_{f_n(x)} \quad \text{and} \quad \hat{B}_n(x) V_x = H_{f_n(x)} \quad \text{for all } x \in Z_n. \tag{9.8}$$

Let m be the probability measure defined on $M \times \mathbb{PR}^2$ by

$$m_n(G) = \frac{1}{2}\mu_n(\{x \in Z_n : V_x \in G\}) + \frac{1}{2}\mu_n(\{x \in Z_n : H_x \in G\}).$$

In other words, m_n projects down to μ_n and its disintegration is given by

$$x \mapsto \frac{1}{2}(\delta_{H_x} + \delta_{V_x}).$$

It is clear from (9.8) that m_n is $\hat{B}_n$-invariant. Using Lemma 5.25 we get that m_n is a linear combination of the probability measures m_n^s and m_n^u defined on $M \times \mathbb{PR}^2$ by

$$m_n^s(G) = \mu_n\big(\{x \in Z_n : (x, E_x^s) \in G\}\big)$$
$$m_n^u(G) = \mu_n\big(\{x \in Z_n : (x, E_x^u) \in G\}\big).$$

Lemma 9.25 *The probability measure m_n is ergodic.*

Proof Suppose that there is an invariant set $\mathscr{X} \subset M \times \mathbb{PR}^2$ with $m_n(\mathscr{X}) \in (0,1)$. Let X_0 be the set of $x \in Z_n$ whose fiber $\mathscr{X} \cap (\{x\} \times \mathbb{PR}^2)$ contains neither H_x nor V_x. In view of (9.8), X_0 is an (f_n, μ_n)-invariant set and so its μ_n-measure is either 0 or 1. Since $m_n(\mathscr{X}) > 0$, we must have $\mu_n(X_0) = 0$. The same kind of argument shows that $\mu_n(X_2) = 0$, where X_2 is the set of $x \in Z_n$ whose fiber contains both H_x and V_x. Now let X_H be the set of $x \in Z_n$ whose fiber contains H_x but not V_x, and let X_V be the set of $x \in Z_n$ whose fiber contains V_x but not H_x. The previous observations show that $X_H \cup X_V$ has full μ_n-measure and it follows from (9.8) that

$$f_n(X_H) = X_V \quad \text{and} \quad f_n(X_V) = X_H.$$

Thus, $\mu_n(X_H) = 1/2 = \mu_n(X_V)$ and $f_n^2(X_H) = X_H$ and $f_n^2(X_V) = X_V$. This is a contradiction because f_n is Bernoulli and, in particular, the second iterate f_n^2 is ergodic. □

Thus, m_n must coincide with either m_n^s and m_n^u. This is a contradiction, because the conditional probabilities of m_n are supported on exactly two points on each fiber, whereas the conditional probabilities of either m_n^u and m_n^s are Dirac masses on a single point. This contradiction proves that the Lyapunov exponents of $\hat{B}_n$ vanish for every n, and that concludes the proof of Theorem 9.22.

9.4 Notes

The original announcement of Theorem 9.16 was made by Mañé in his invited address to the ICM 1983, in Warsaw [88]. A draft of the proof circulated for several years, but it remained incomplete when Mañé passed away in 1995. The first complete proof was provided by Bochi in his thesis, based on that draft, and was published in [30].

A few related results had been obtained in the meantime. Knill [75, 76] considered L^∞ cocycles with values in $\mathrm{SL}(2)$ and proved that, as long as the base dynamics is aperiodic, the set of cocycles with non-zero exponents is never open. This was refined to the C^0 case by Bochi [27], who proved that an $\mathrm{SL}(2)$-cocycle is a continuity point for the Lyapunov exponents in $C^0(M, \mathrm{SL}(2))$ if and only if it is hyperbolic or else the exponents vanish (Corollary 9.8).

Our presentation in Section 9.2 is adapted from the texts of Bochi [27] and Avila and Bochi [9]. A different strategy was used by Bochi [30], extending to the special case of derivative cocycles.

Related results were also obtained for L^p cocycles with $p < \infty$. Arbieto and Bochi [4] proved that Lyapunov exponents are still semi-continuous on the cocycle relative to the L^p-norm, for any $1 \le p < \infty$. Then, using a result of Arnold and Cong [6], they concluded that the cocycles whose exponents are all equal are precisely the continuity points for the Lyapunov exponents on

$L^p(\mu)$. It follows that these cocycles form a residual subset of $L^p(\mu)$. See also Bessa and Vilarinho [25].

Bochi and Viana [34] extended Theorems 9.5 and 9.16 to arbitrary dimension (Theorems 9.18 and 9.19). They also proved a version of Theorem 9.19 for symplectic diffeomorphisms which was later improved by Bochi [31] to obtain Theorem 9.21. Moreover, Bessa and his coauthors [3, 20, 21, 22, 23, 24] extended most of these statements to the continuous-time setting, as we explained at the end of Section 9.2.4.

The examples in Section 9.3 were constructed by Bocker and Viana [35], based on the method of proof of Theorem 9.5. Such examples notwithstanding, it should be stressed that the behavior of Hölder-continuous cocycles is usually very different from the behavior of typical continuous cocycles. Example 9.7 provides a striking illustration of this fact: one can use the invariance principle to show (see [36, Corollary 12.34]) that, under two mild additional assumptions, every Hölder-continuous cocycle in a C^0-neighborhood has *non-vanishing* Lyapunov exponents. See Chapter 12 of Bonatti, Díaz and Viana [36] and the survey papers of Bochi and Viana [32, 33] for more detailed discussions.

9.5 Exercises

Exercise 9.1 Prove that the maps $A \mapsto \Lambda^k A$, $1 \le k \le d-1$ are continuous for the topology of uniform convergence (C^0-norm). Moreover, the same is true for the L^1-norm, restricted to maps $A : M \to \mathrm{GL}(d)$ with $\|A\|_\infty \le C$, for any constant $C > 1$.

Exercise 9.2 Let $A : M \to \mathrm{GL}(d)$ be such that $\|A^{\pm 1}\| \in L^\infty(\mu)$. Show that for every $\varepsilon > 0$ and $C > 1$ there exists $\delta > 0$ such that

$$\lambda_+(B,\mu) \le \lambda_+(A,\mu) + \varepsilon \quad \text{and} \quad \lambda_-(B,\mu) \le \lambda_-(A,\mu) - \varepsilon$$

for any $B : M \to \mathrm{GL}(d)$ with $\|B^{\pm 1}\|_\infty \le C$ and $\|A - B\|_1 \le \delta$. Moreover, this statement remains true if one replaces λ_+ and λ_- by, respectively, $\sum_{i=1}^k \chi_i$ and $\sum_{i=d-k+1}^d \chi_i$, for any $0 \le k \le d$.

Exercise 9.3 Let X be a complete metric space, F be a closed subset and A_n, $n \ge 1$ be open subsets whose closure contains F for every n. Show that the closure of $\bigcap_n A_n$ contains F.

Exercise 9.4 Use the theorem of Lusin and the Tietze extension theorem to show that, given any measurable function $B \in \mathrm{SL}(d)$ and any $\rho > 0$, there exist continuous functions $J : M \to \mathrm{SL}(d)$ such that $\|J\|_\infty \leq \|B\|_\infty$ and $J = B$ outside a set with measure ρ.

Exercise 9.5 Suppose that the linear cocycle F admits a dominated decomposition $\mathbb{R}^d = V_x^1 \oplus \cdots \oplus V_x^l$, $x \in \Lambda$ on some invariant set Λ. Conclude that:

(1) The decomposition is continuous (the subspaces V_x^i depend continuously on $x \in \Lambda$), and even admits a continuous extension to the closure of Λ.
(2) Any linear cocycle sufficiently close to F in the uniform convergence norm admits a dominated decomposition on Λ into subspaces with the same dimensions.

Exercise 9.6 Show that an $\mathrm{SL}(2)$-cocycle admits a dominated decomposition over an invariant set $\Lambda \subset M$ if and only if it is hyperbolic over Λ.

Exercise 9.7 Prove the following fact, which was implicit in Section 9.3: If Z is a cylinder in a Bernoulli shift space (M, μ) then Z is not a coboundary set.

10
Continuity

We have seen in Chapter 9 that the Lyapunov exponents may depend in a complicated way on the underlying linear cocycle. The theme, in the context of products of random matrices, of the present chapter is that this dependence is always continuous.

Let $\mathscr{G}(d)$ denote the space of compactly supported probability measures p on $\mathrm{GL}(d)$, endowed with the following topology: p' is close to p if it is close in the weak*-topology and $\operatorname{supp} p'$ is contained in a small neighborhood of $\operatorname{supp} p$. Let $\lambda_+(p)$ and $\lambda_-(p)$ denote the extremal Lyapunov exponents of the product of random matrices associated with a given $p \in \mathscr{G}(d)$, in the sense of Section 2.1.1. In other words, $\lambda_\pm(p) = \lambda_\pm(A,\mu)$, where $A : \mathrm{GL}(d)^{\mathbb{N}} \to \mathrm{GL}(d)$, $(\alpha_k)_k \mapsto \alpha_0$ and $\mu = p^{\mathbb{N}}$. A probability measure η on $\mathbb{PR}^d$ will be called *p-stationary* if it is stationary for this cocycle. We are going to prove:

Theorem 10.1 (Bocker and Viana) *The functions $\mathscr{G}(2) \to \mathbb{R}$, $p \mapsto \lambda_\pm(p)$ are continuous at every point in the domain.*

Avila, Eskin and Viana announced recently that this statement remains true in arbitrary dimension. Even more, for any $d \geq 2$, all the Lyapunov exponents depend continuously on the probability distribution $p \in \mathscr{G}(d)$. The proof will appear in [12].

It is easy to see that Theorem 10.1 implies Theorem 1.3. Indeed, given $(A_{i,j}, p_j)_{i,j} \in \mathrm{GL}(2)^m \times \Delta^m$, let us consider $p = \sum_j p_j \delta_{A_j} \in \mathscr{G}(2)$. For any nearby element $(A'_{i,j}, p'_j)_{i,j}$ of $\mathrm{GL}(2)^m \times \Delta^m$, the corresponding probability measure $p' = \sum_j p'_j \delta_{A'_j}$ is close to p inside $\mathscr{G}(2)$; note that the assumption that $p_j > 0$ for every j is crucial for ensuring that $\operatorname{supp} p'$ is contained in a small neighborhood of $\operatorname{supp} p$. So, the Lyapunov exponents $\lambda_\pm(A'_{i,j}, p'_j)_{i,j} = \lambda_\pm(p')$ are close to $\lambda_\pm(A_{i,j}, p_j)_{i,j} = \lambda_\pm(p)$, as claimed in Theorem 1.3.

Here is another direct consequence of Theorem 10.1. Let X be a separable complete metric space and $f : M \to M$ be the shift map on $M = X^{\mathbb{N}}$. Consider the space of measurable functions $A : X \to \mathrm{GL}(2)$ such that $\log \|A^{\pm 1}\|$ are bounded, with the topology of uniform convergence. Fix any probability measure p on X and let $\lambda_\pm(A,p)$ be the extremal Lyapunov exponents relative to $\mu = p^{\mathbb{N}}$ of the locally constant linear cocycle defined by A over f. Observe that $\lambda_\pm(A,p) = \lambda_\pm(A_*p)$ and $A \mapsto A_*p$ is continuous. Thus, it follows from Theorem 10.1 that the functions $A \mapsto \lambda_\pm(A,p)$ are continuous at every point.

The proof of Theorem 10.1 occupies Sections 10.1 through 10.5. Then, in Section 10.6, we try to put this theorem and the results of Chapter 9 together in a consistent possible scenario.

10.1 Invariant subspaces

According to Corollary 9.3, we only need to consider the case $\lambda_-(p) < \lambda_+(p)$. Let $(p_k)_k$ be a sequence converging to some p in $\mathscr{G}(2)$. In other words, given any continuous functions $\varphi_1, \dots, \varphi_l : \mathrm{GL}(d) \to \mathbb{R}$ and any $\varepsilon > 0$,

$$|\int \varphi_j \, dp_k - \int \varphi_j \, dp| < \varepsilon \text{ for } j = 1, \dots, l$$

and $\operatorname{supp} p_k$ is contained in the ε-neighborhood of $\operatorname{supp} p$, for every large k. We are going to show that $\lambda_+(p_k) \to \lambda_+(p)$ as $k \to \infty$. Then, by Exercise 10.1, we also have $\lambda_-(p_k) \to \lambda_-(p)$ as $k \to \infty$.

For each k, let η_k be a p_k-stationary measure that realizes the largest Lyapunov exponent for p_k:

$$\lambda_+(p_k) = \int \Phi \, d\eta_k \, dp_k.$$

Up to restricting to a subsequence, we may suppose that $(\eta_k)_k$ converges to some probability measure η on $\mathbb{PR}^2$. By Proposition 5.9(b), this is a p-stationary measure. Moreover, since Φ is continuous,

$$\lambda_+(p_k) = \int \Phi \, d\eta_k \, dp_k \quad \text{converges to} \quad \int \Phi \, d\eta \, dp.$$

Thus, if η realizes the largest Lyapunov exponent for p then we are done. In what follows we suppose that

$$\int \Phi \, d\eta \, dp < \lambda_+(p) \tag{10.1}$$

and we prove that this leads to a contradiction.

We will use $\underline{\alpha} = (\alpha_n)_n$ to represent a generic point of $M = \mathrm{GL}(d)^{\mathbb{N}}$ and we will write $\alpha^{(n)} = \alpha_{n-1} \cdots \alpha_0$ for $n \geq 1$.

Proposition 10.2 *There exists some $L \in \mathbb{PR}^2$ such that*

(a) $\eta(\{L\}) > 0$;
(b) $\alpha(L) = L$ *for every* $\alpha \in \operatorname{supp} p$;
(c) *for* μ*-almost all* $\underline{\alpha} \in \mathrm{GL}(2)^{\mathbb{N}}$,

$$\lim_n \frac{1}{n} \log \|\alpha^{(n)}(v)\| = \begin{cases} \lambda_-(p) & \text{if } v \in L \setminus \{0\} \\ \lambda_+(p) & \text{if } v \in \mathbb{R}^2 \setminus L. \end{cases}$$

Proof According to Proposition 5.5, the product measure $\mu \times \eta$ is invariant under the projective cocycle $\mathbb{P}F : M \times \mathbb{PR}^2 \to M \times \mathbb{PR}^2$ defined by the function $A : M \to \mathrm{GL}(2)$ over the shift $f : M \to M$ on $M = \mathrm{GL}(2)^{\mathbb{N}}$. So, we may use the ergodic theorem to conclude that

$$\tilde{\Phi}(\underline{\alpha}, [v]) = \lim_n \sum_{j=0}^{n-1} \Phi(\mathbb{P}F^j(\underline{\alpha}, [v])) = \lim_n \frac{1}{n} \log \|\alpha^{(n)}(v)\|$$

exists for $\mu \times \eta$-almost every point, is constant on the orbits of $\mathbb{P}F$ and satisfies $\int \tilde{\Phi} \, d\eta \, d\mu = \int \Phi \, d\eta \, d\mu$. Thus, the assumption implies that $\tilde{\Phi} < \lambda_+(p)$ on some subset with positive measure for $\mu \times \eta$.

We claim that for μ-almost every $\underline{\alpha} \in M$ there exists a unique $L(\underline{\alpha}) \in \mathbb{PR}^2$ such that $\tilde{\Phi}(\underline{\alpha}, L(\underline{\alpha})) < \lambda_+(p)$. Moreover,

$$L(f(\underline{\alpha})) = \alpha_0(L(\underline{\alpha})) \quad \text{for } \mu\text{-almost every } \underline{\alpha} \in M. \tag{10.2}$$

Indeed, consider

$$\mathscr{Z} = \{(\underline{\alpha}, [v]) \in M \times \mathbb{PR}^2 : \tilde{\Phi}(\underline{\alpha}, [v]) < \lambda_+(p)\}.$$

This is a measurable, $\mathbb{P}F$-invariant set with positive measure for $\mu \times \eta$. Hence, the projection $Z = \{\underline{\alpha} \in M : (\underline{\alpha}, [v]) \in \mathscr{Z} \text{ for some } [v]\}$ is a measurable (Proposition 4.5), f-invariant set with positive measure for μ. Since (f, μ) is ergodic, it follows that $\mu(Z) = 1$. This proves the existence part of the claim.

Next, given any $\underline{\alpha} \in M$, suppose that there exist two distinct points $[v_1]$ and $[v_2]$ in projective space such that $\tilde{\Phi}(\underline{\alpha}, [v_i]) < \lambda_+(p)$ for $i = 1, 2$. Since every vector in $\mathbb{R}^2$ may be written as linear combination of v_1 and v_2, it follows that

$$\lim_n \frac{1}{n} \log \|\alpha^{(n)}\| < \lambda_+(p).$$

This can only happen on a subset with zero measure for μ, because the largest Lyapunov exponent is equal to $\lambda_+(p)$ at μ-almost every point $\underline{\alpha} \in M$. That proves the uniqueness part of the claim. Finally, (10.2) is a direct consequence of the fact that the function $\tilde{\Phi}$ is constant on the orbits of $\mathbb{P}F$.

Let η_0 be any ergodic component (Theorem 5.14) of η such that $\mu \times \eta_0(\mathscr{Z})$ is positive. Then $\mu \times \eta_0(\mathscr{Z}) = 1$, by ergodicity, and so $\eta_0(\{L(\underline{\alpha})\}) = 1$ for μ-almost every $\underline{\alpha} \in M$. It follows that there exists $L \in \mathbb{PR}^2$ such that $L(\underline{\alpha}) =$

L for μ-almost every $\underline{\alpha} \in M$. This subspace L satisfies the conditions in the statement. Indeed, condition (a) is given by $\eta(\{L\}) = \mu \times \eta(\mathscr{L}) > 0$. The relation (10.2) gives that $\alpha(L) = L$ for p-almost every α, which is equivalent to condition (b). By construction, $\tilde{\Phi}(\underline{\alpha}, [v]) = \lambda_+(p)$ for $[v] \neq L$ and $\tilde{\Phi}(\underline{\alpha}, L) < \lambda_+(p)$ for μ-almost all $\underline{\alpha} \in M$. Since $\tilde{\Phi}$ only takes the values $\lambda_\pm(p)$, by the Oseledets theorem, it follows that $\tilde{\Phi}(\underline{\alpha}, L) = \lambda_-(p)$ for μ-almost all $\underline{\alpha} \in M$. Thus, L satisfies condition (c) as well. □

10.2 Expanding points in projective space

Let L be the subspace given by Proposition 10.2. We are going to view L as a kind of *repelling fixed point* for the random walk defined by p on the projective space and to analyse such a repeller from the point of view of (the random walks defined by) nearby probability measures. The first step will be to establish in which sense L is repelling. For that, we need to introduce some notation. Most of the time the dimension d will be arbitrary, although our interest is in $d = 2$.

Let $\mathscr{P}(d)$ be the space of probability measures on $\mathbb{PR}^d$, endowed with the weak* topology. Given $p \in \mathscr{G}(d)$ and $\eta \in \mathscr{P}(d)$, we denote by $p * \eta$ the push-forward of the product measure $p \times \eta$ under the map $(\alpha, x) \mapsto \alpha(x)$. That is,

$$p * \eta(D) = \int \eta(\alpha^{-1}(D))\, dp(\alpha) \quad \text{for every measurable set } D \subset \mathbb{PR}^d. \tag{10.3}$$

Observe that a probability η is p-stationary if and only if $p * \eta = \eta$.

Similarly, given $p_1, p_2 \in \mathscr{G}(d)$, we denote by $p_1 * p_2 \in \mathscr{G}(d)$ the push-forward of the product measure $p_1 \times p_2$ under the map $(\alpha_1, \alpha_2) \mapsto \alpha_1 \alpha_2$. In other words,

$$p_1 * p_2(E) = \int p_2(\alpha_1^{-1} E)\, dp_1(\alpha_1) = \int p_1(E \alpha_2^{-1})\, dp_2(\alpha_2) \tag{10.4}$$

for every measurable set $E \subset \mathrm{GL}(d)$. We call $p_1 * p_2$ the *convolution* of p_1 and p_2. Clearly, $p_1 * p_2$ is compactly supported if p_1 and p_2 are. Given $p \in \mathscr{G}(d)$, define $p^{(1)} = p$ and $p^{(n+1)} = p * p^{(n)}$ for every $n \geq 1$.

Remark 10.3 The operations (10.3) and (10.4) make sense, more generally, for measures (not necessarily probabilities) p_1, p_2 and p in $\mathrm{GL}(d)$ and η in $\mathbb{PR}^d$.

Given $\alpha \in \mathrm{GL}(d)$ and $[v] \in \mathbb{PR}^d$, we denote by $D\alpha([v])$ the derivative at the point $[v]$ of the action of α in the projective space. In explicit terms (see

Exercise 10.3):

$$D\alpha([v])\dot{v} = \frac{\mathrm{proj}_{\alpha(v)}\,\alpha(\dot{v})}{\|\alpha(v)\|/\|v\|} \quad \text{for every } \dot{v} \in T_{[v]}\mathbb{P}\mathbb{R}^d = \{v\}^\perp \tag{10.5}$$

where $\mathrm{proj}_w : u \mapsto u - w(u\cdot w)/(w\cdot w)$ denotes the orthogonal projection to the hyperplane orthogonal to w. When $d=2$, the tangent hyperplanes are lines and so $D\alpha([v])$ is a real number.

Given $p \in \mathscr{G}(d)$ and $x \in \mathbb{P}\mathbb{R}^d$, we say that x is *p-invariant* if it is a fixed point for p-almost every $\alpha \in \mathrm{GL}(d)$ or, equivalently, for every $\alpha \in \mathrm{supp}\, p$. Then we say that x is *p-expanding* if there exist $\ell \geq 1$ and $c > 0$ such that

$$\int \log \|D\beta(x)\dot{v}\|\, dp^{(\ell)}(\beta) \geq 2c \quad \text{for every unit vector } \dot{v} \in T_x\mathbb{P}\mathbb{R}^d. \tag{10.6}$$

Part (b) of Proposition 10.2 means that L is p-invariant. In view of part (c), the next proposition implies that L is also p-expanding:

Proposition 10.4 *Let $x \in \mathbb{P}\mathbb{R}^d$ be a p-invariant point and suppose that there exist $a < b$ such that*

$$\lim_n \frac{1}{n} \log \|\alpha^{(n)}(v)\| \begin{cases} \leq a & \text{if } v \in x\setminus\{0\} \\ \geq b & \text{if } v \in \mathbb{R}^d \setminus x. \end{cases}$$

for μ-almost all $\underline{\alpha} \in M$. Then x is a p-expanding point.

Proof By Proposition 4.14 and the hypothesis,

$$\lim_n \frac{1}{n} \log \|\mathrm{proj}_{\alpha^{(n)}(x)}\, \alpha^{(n)}(\dot{v})\| = \lim_n \frac{1}{n} \log \|\alpha^{(n)}(\dot{v})\| \geq b$$

for every unit vector $\dot{v} \in T_x\mathbb{R}\mathbb{P}^d = \{x\}^\perp$ and μ-almost every $\underline{\alpha} \in M$. Then,

$$\lim_n \frac{1}{n} \log \|D\alpha^{(n)}(x)\dot{v}\| \geq b - a > 0$$

for every unit vector $\dot{v} \in \{x\}^\perp$ and μ-almost every $\underline{\alpha} \in M$. Since the support of p is compact, the inequalities in Exercise 10.3 ensure that there exists $C > 0$ such that $-C \leq \log \|D\alpha(x)\dot{v}\| \leq C$ for every $\alpha \in \mathrm{supp}\, p$ and every univ vector $\dot{v} \in \{v\}^\perp$. It follows that

$$-C \leq \frac{1}{n} \log \|D\alpha^{(n)}(x)\dot{v}\| \leq C$$

for every $\underline{\alpha} \in \mathrm{supp}\, p^{(n)}$, every unit vector $\dot{v} \in \{v\}^\perp$ and every $n \geq 1$. So, using the bounded convergence theorem, the previous inequality implies that

$$\lim_n \int \frac{1}{n} \log \|D\beta(x)\dot{v}\|\, dp^{(n)}(\beta) = \lim_n \int \frac{1}{n} \log \|D\alpha^{(n)}(x)\dot{v}\|\, d\mu(\underline{\alpha}) \geq b - a$$

for every unit vector $\dot{v} \in \{x\}^{\perp}$. Let $c = (b-a)/2$ and, for each unit vector $\dot{v} \in \{x\}^{\perp}$, let $n(\dot{v}) \geq 1$ be the smallest value of n such that

$$\int \frac{1}{n} \log \|D\beta(x)\dot{v}\| dp^{(n)}(\beta) > c.$$

Note that $n(\dot{v})$ depends upper semi-continuously on $\dot{v}$, and so

$$n_0 = \sup\{n(\dot{v}) : \dot{v} \in \{x\}^{\perp} \text{ with } \|\dot{v}\| = 1\}$$

is finite. Fix $C > 0$ such that

$$\int \log \|D\beta(x)\dot{v}\| dp^{(n)}(\beta) \geq -C$$

for $1 \leq n \leq n_0$ and every unit vector $\dot{v} \in \{x\}^{\perp}$. Then

$$\int \log \|D\beta(x)\dot{v}\| dp^{(n)}(\beta) \geq c\left[\frac{n}{n_0}\right] n_0 - C \geq 2c$$

for every unit vector $\dot{v} \in \{x\}^{\perp}$, if n is sufficiently large. □

The heart of the proof of Theorem 10.1 is the statement that *p-expanding points are invisible for the p-stationary measures that are limits of p_k-stationary non-atomic measures with $p_k \to p$*. In precise terms:

Theorem 10.5 *Suppose that $x \in \mathbb{PR}^d$ is a p-expanding point. Let $(p_k)_k$ be a sequence in $\mathcal{G}(d)$ converging to p and, for each $k \geq 1$, let $\eta_k \in \mathcal{P}(d)$ be a p_k-stationary measure. Suppose that $(\eta_k)_k$ converges to some η. If η_k is non-atomic for k arbitrarily large, then $\eta(\{x\}) = 0$.*

The proof of this theorem will be given in Sections 10.4 and 10.5. Right now, let us explain how it can be used to complete the proof of Theorem 10.1.

10.3 Proof of the continuity theorem

The existence of an invariant subspace L implies that no matrix $\alpha \in \operatorname{supp} p$ is elliptic. If the support consisted only of parabolic matrices (with the same invariant subspace) and multiples of the identity then we would have $\lambda_-(p) = \lambda_+(p)$, which is assumed not to be the case. Thus, $\operatorname{supp} p$ contains some hyperbolic matrix. Since the set of hyperbolic matrices is open, it follows that it has positive measure for p. This fact will be useful in a while.

By Proposition 10.2 and Theorem 10.5, we may suppose that η_k is atomic for every large k. Then, by Lemma 6.9, there exists a finite set $\mathcal{L}_k \subset \mathbb{PR}^2$ such that $\alpha(\mathcal{L}_k) = \mathcal{L}_k$ for every $\alpha \in \operatorname{supp} p$ and $\eta(\{y\}) > 0$ for every $y \in \mathcal{L}_k$. The open set of hyperbolic matrices has positive measure for p_k, for every large k, because p_k is close to p in the weak* topology. Observe, furthermore, that for a hyperbolic matrix any finite invariant set consists of either one or two eigenspaces. Thus, for every large k, the set $\mathscr{L}_k$ can have no more than 2 elements.

First, suppose that $\mathscr{L}_k$ has a single element L_k. We claim that

$$\int \Phi(\alpha, L_k)\, dp_k(\alpha) = \lambda_+(p_k). \tag{10.7}$$

To prove this, let ζ_k be the Dirac mass at the point L_k. The product measure $\mu_k \times \zeta_k$ is $\mathbb{P}F$-invariant, and so it follows from the ergodic theorem that

$$\int \Phi(\alpha, L_k)\, dp_k(\alpha) = \int \tilde{\Phi}(\underline{\alpha}, L_k)\, d\mu_k(\underline{\alpha}). \tag{10.8}$$

We also have that $\tilde{\Phi}(\underline{\alpha}, [v]) = \lambda_+(p_k)$ for $\mu_k \times \eta_k$-almost every $(\underline{\alpha}, [v])$, because $\int \tilde{\Phi}(\underline{\alpha}, [v])\, d\eta_k([v])\, d\mu_k(\underline{\alpha}) = \lambda_+(p_k)$ and $\tilde{\Phi}(\underline{\alpha}, [v]) \le \lambda_+(p_k)$ for μ_k-almost every $\underline{\alpha} \in M$ and every $[v] \in \mathbb{P}\mathbb{R}^2$. Since $\eta_k(\{L_k\}) > 0$, this implies $\tilde{\Phi}(\underline{\alpha}, L_k) = \lambda_+(p_k)$ for μ_k-almost every $\underline{\alpha} \in M$. Together with (10.8), this implies the claim (10.7).

Up to restricting to a subsequence, may assume that $(L_k)_k$ converges to some subspace $L_0 \in \mathbb{P}\mathbb{R}^2$. Let ζ_0 be the Dirac mass at L_0. Since $\operatorname{supp} p_k$ converges to $\operatorname{supp} p$ in the Hausdorff distance (Exercise 10.4), it follows that $\alpha(L_0) = L_0$ for every $\alpha \in \operatorname{supp} p$. Thus, the product measure $\mu \times \zeta_0$ is invariant under the projective cocycle $\mathbb{P}F$. Since Φ is a continuous function, passing to the limit in (10.7) we get that

$$\lambda_+(p_k) \to \int \Phi(\alpha, L_0)\, dp(\alpha) = \int \tilde{\Phi}(\underline{\alpha}, L_0)\, d\mu(\underline{\alpha}) \tag{10.9}$$

According to Proposition 10.2, there are two possibilities:

(i) If $L_0 \neq L$ then the right-hand side of (10.9) is equal to $\lambda_+(p)$. Then $\int \Phi\, d\eta_k\, d\mu_k = \lambda_+(p_k)$ converges to $\lambda_+(p)$, which contradicts (10.1).
(ii) If $L_0 = L$ then the right-hand side of (10.9) is $\lambda_-(p)$. Then (Exercise 10.1) we also have that $\lambda_-(p_k) \to \lambda_+(p)$. This is a contradiction, because $\lambda_-(p_k) \le \lambda_+(p_k)$ for every k and $\lambda_-(p) < \lambda_+(p)$.

We have shown that, for k large, $\mathscr{L}_k$ cannot consist of a single point.

Finally, suppose that, for all k large, $\mathscr{L}_k$ consists of two points, L_k and L'_k, with positive measure for η. Let $\zeta_k = (\delta_{L_k} + \delta_{L'_k})/2$. Arguing as we did for (10.7), we find that

$$\frac{1}{2}\int \Phi(\alpha, L_k)\, dp_k(\alpha) + \frac{1}{2}\int \Phi(\alpha, L'_k)\, dp_k(\alpha) = \lambda_+(p_k). \tag{10.10}$$

We may assume that $(L_k)_k$ and $(L'_k)_k$ converge to subspaces L_0 and L'_0, respectively. Let $\zeta_0 = (\delta_{L_0} + \delta_{L'_0})/2$. Passing to the limit in (10.10),

$$\begin{aligned}\lambda_+(p_k) \to &\frac{1}{2}\int \Phi(\alpha, L_0)\,dp(\alpha) + \frac{1}{2}\int \Phi(\alpha, L'_0)\,dp(\alpha)\\ &= \frac{1}{2}\int \tilde{\Phi}(\alpha, L_0)\,dp(\alpha) + \frac{1}{2}\int \tilde{\Phi}(\alpha, L'_0)\,dp(\alpha).\end{aligned} \tag{10.11}$$

If L_0 and L'_0 are both different from L then the right-hand side is equal to $\lambda_+(p)$ and we reach a contradiction just as we did in case (i) of the previous paragraph. Now suppose that $L_0 = L$. We may assume that, for each k large, there exists $\alpha_k \in \operatorname{supp} p_k$ such that $\alpha_k(L_k) = L'_k$: otherwise, we could take $\mathscr{L}_k = \{L_k\}$ instead, and that case has already been dealt with. Passing to the limit along a convenient subsequence, we find $\alpha_0 \in \operatorname{supp} p$ such that $\alpha_0(L_0) = L'_0$. By part (b) of Proposition 10.2, this implies that L'_0 is also equal to L. Then the right-hand side of (10.11) is equal to $\lambda_-(p)$, and we reach a contradiction just as we did in case (ii) of the previous paragraph. So, $\mathscr{L}_k$ cannot consist of two points either.

We have reduced the proof of Theorem 10.1 to proving Theorem 10.5.

10.4 Couplings and energy

For simplicity, we will write $P = \mathbb{PR}^d$ and $G = \mathrm{GL}(d)$. Let d be the distance on P defined by the angle between two directions. For any Borel measure ξ on $P \times P$ and $\delta > 0$, define the δ-*energy* of ξ to be

$$E_\delta(\xi) = \int d(x,y)^{-\delta}\,d\xi(x,y).$$

Let $\pi^i : P \times P \to P$ be the projection on the ith coordinate, for $i = 1, 2$.

By definition, the *mass* of a measure η on P is the total measure $\|\eta\| = \eta(P)$ of the ambient space. If η_1 and η_2 are measures on P with the same mass, a *coupling* of η_1 and η_2 is a measure on $P \times P$ that projects to η_i on the ith coordinate for $i = 1, 2$. For instance,

$$\xi = \frac{1}{\|\eta_1\|}\,\eta_1 \times \eta_2 = \frac{1}{\|\eta_2\|}\,\eta_1 \times \eta_2$$

is a coupling of η_1 and η_2. Define:

$$e_\delta(\eta_1, \eta_2) = \inf\{E_\delta(\xi) : \xi \text{ a coupling of } \eta_1 \text{ and } \eta_2\}. \tag{10.12}$$

The infimum is always achieved, because the function $\xi \mapsto E_\delta(\xi)$ is lower semi-continuous (Exercise 10.5).

A *self-coupling* of a measure $\eta \in \mathscr{P}(d)$ is a coupling of η and η. We call a self-coupling *symmetric* if it is invariant under the involution $\iota : (x,y) \mapsto (y,x)$. Define the δ*-energy* of η to be:

$$e_\delta(\eta) = e_\delta(\eta,\eta) = \inf\{E_\delta(\xi) : \xi \text{ a self-coupling of } \eta\}. \tag{10.13}$$

We call $p * \eta$ the *convolution* of η by p. Observe that a probability η is p-stationary if and only if $p * \eta = \eta$. If ξ is any self-coupling then $\xi' = (\xi + \iota_* \xi)/2$ is a symmetric self-coupling and $E_\delta(\xi) = E_\delta(\xi')$ Thus, the infimum in (10.13) is always achieved at some symmetric self-coupling of η. Such symmetric self-couplings are called δ*-optimal.*

Example 10.6 Let $P = [0,2]$ and η be the convex combination of the Dirac masses at 0, 1 and 2; that is,

$$\eta = \frac{1}{3}\delta_0 + \frac{1}{3}\delta_1 + \frac{1}{3}\delta_2.$$

Since η is a probability, the product $\xi_1 = \eta \times \eta$ is a (symmetric) self-coupling of η. Observe that the product ξ_1 has atoms on the diagonal of $P \times P$ and, thus, its δ-energy is infinite. Now, consider

$$\xi_2 = \frac{1}{6}\delta_{(0,1)} + \frac{1}{6}\delta_{(0,2)} + \frac{1}{6}\delta_{(1,0)} + \frac{1}{6}\delta_{(1,2)} + \frac{1}{6}\delta_{(2,0)} + \frac{1}{6}\delta_{(2,1)}.$$

Observe that ξ_2 is a symmetric self-coupling of η. Moreover, the support of ξ_2 does not intersect the diagonal and, consequently, $E_\delta(\xi_2)$ is finite. This proves that $e_\delta(\eta) < \infty$. It is easy to check that ξ_2 is a δ-optimal self-coupling of η.

This notion of energy allows us to characterize the presence of atoms:

Lemma 10.7 *For any $\eta \in \mathscr{P}(d)$ and $\delta > 0$:*

(1) *if $\eta(\{x\}) > \|\eta\|/2$ for some $x \in P$ then $e_\delta(\eta) = \infty$;*
(2) *if $\eta(\{x\}) < \|\eta\|/2$ for every $x \in P$ then $e_\delta(\eta) < \infty$.*

Proof First we prove (1). Take x as in the hypothesis. For any self-coupling ξ of η, we have

$$\xi(\{x\}^c \times P) = \eta(\{x\}^c) = \xi(P \times \{x\}^c).$$

Taking the union, $\xi(\{(x,x)\}^c) \leq 2\eta(\{x\}^c) < \eta(P) = \xi(P \times P)$. This implies that (x,x) is an atom for ξ, and so $E_\delta(\xi) = \infty$.

Now we prove (2). The hypothesis implies (Exercise 10.7) that there exists $\rho > 0$ such that $\eta(B(x,2\rho)) < \eta(B(x,2\rho)^c)$ for every $x \in P$. Let $x_1, \ldots, x_l$ be such that $\mathscr{U} = \{B(x_i, \rho) : i = 1, \dots, l\}$ covers P. We claim that there exists a finite sequence $\xi_0, \dots, \xi_l$ of symmetric self-couplings of η such that

$$\xi_j\big(B(x_i,\rho) \times B(x_i,\rho)\big) = 0 \quad \text{for every } 1 \le i \le j. \tag{10.14}$$

Denote $\Delta_r = \{(y,z) \in P \times P : d(y,z) < r\}$. Take $r > 0$ to be a Lebesgue number for the open cover $\mathscr{U}$. Then, by definition,

$$\Delta_r \subset \bigcup_{i=1}^{l} B(x_i,\rho) \times B(x_i,\rho).$$

It follows that $\xi_l\big(\Delta_r\big) = 0$ and so $E_\delta(\xi_l) < \infty$. We are left to prove our claim.

We are going to construct the sequence ξ_j by induction of j. Let ξ_0 be an arbitrary self-coupling of η. For each $j = 1, \dots, l$, suppose that ξ_{j-1} has been constructed in such a way that $B(x_i,\rho) \times B(x_i,\rho)$ has zero measure for $1 \le i \le j-1$. Denote $\xi = \xi_{j-1}$ and $B = B(x_j,\rho)$ and $C = B(x_j, 2\rho)$ and $D = B(x_j, 2\rho)^c$. Note that (see Figure 10.1)

- $\xi(C \times C) + \xi(C \times D) = \eta(C)$, because $\pi^1_*\xi = \eta$;
- $\xi(D \times C) + \xi(D \times D) = \eta(D)$, because $\pi^1_*\xi = \eta$;
- $\xi(C \times D) = \xi(D \times C)$, because ξ is symmetric.

Since $\eta(C) < \eta(D)$, this implies that $\xi(B \times B) \le \xi(C \times C) < \xi(D \times D)$. Let $\theta = \xi(B \times B)/\xi(D \times D)$ and then let ζ be any coupling of $\pi^1_*\big(\xi \mid B \times B\big)$ and $\theta\pi^1_*\big(\xi \mid D \times D\big)$. This is well defined, because of the choice of θ. Moreover, ζ is concentrated on $B \times D$ and, in particular, it satisfies $\zeta(\Delta_\rho) = 0$. Define (see Figure 10.1)

$$\xi_j = \xi - \big(\xi \mid B \times B\big) - \theta\big(\xi \mid D \times D\big) + \zeta + \iota_*\zeta. \tag{10.15}$$

Observe that ξ_j is a (positive) measure, because $\theta < 1$. It is also clear that ξ_j is symmetric. Moreover,

$$\begin{aligned}\pi^1_*\xi_j = \pi^1_*\xi - \pi^1_*\big(\xi \mid B \times B\big) - \theta\pi^1_*\big(\xi \mid D \times D\big) \\ + \pi^1_*\big(\xi \mid B \times B\big) + \theta\pi^1_*\big(\xi \mid D \times D\big) = \pi^1_*\xi = \eta.\end{aligned}$$

This shows that ξ_j is indeed a self-coupling of η. The definition (10.15) also gives that $\xi_j(B \times B) = 0$ and (since ζ and $\iota_*\zeta$ give zero weight to Δ_ρ)

$$\xi_j\big(B(x_i,\rho) \times B(x_i,\rho)\big) \le \xi\big(B(x_i,\rho) \times B(x_i,\rho)\big) = 0$$

for every $1 \le i \le j-1$. Thus, ξ_j satisfies (10.14). □

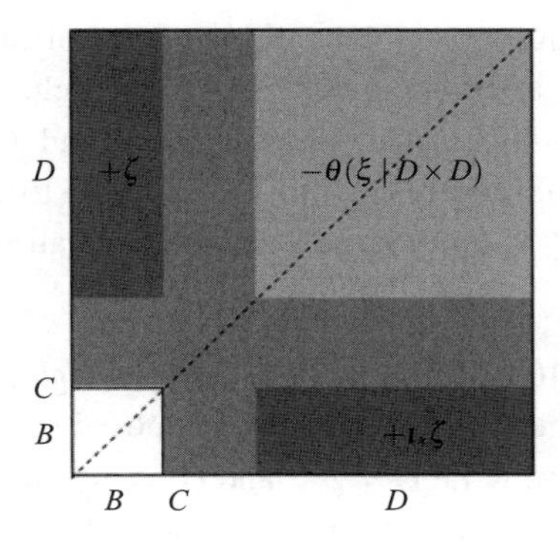

Figure 10.1 Removing mass from the neighborhood of the diagonal

10.5 Conclusion of the proof

We are ready to finish the proof of Theorem 10.5 and, thus, of Theorem 10.1. Keep in mind that we denote $P = \mathbb{PR}^d$ and $G = \mathrm{GL}(d)$.

Let x be a p-expanding point. Fix constants $c > 0$ and $\ell \geq 1$ as in the definition (10.6). Up to restricting to a subsequence, we may suppose that $(\eta_k)_n$ converges to η. Assume, for contradiction, that $\eta(\{x\}) > 0$. Then let $U_0 \subset P$ be an open neighborhood of x such that

$$\eta(\{x\}) > \frac{9}{10}\eta(\bar{U}_0).$$

Lemma 10.8 *Assuming that $\delta > 0$ is small enough and $k \geq 1$ is large enough, then $\int \|D\beta(x)\dot{v}\|^{-\delta}\, dp_k^{(\ell)}(\beta) \leq 1 - c\delta$ for every unit vector $\dot{v} \in T_xP$.*

Proof Fix some compact neighborhood V of the support of p. Take k to be large enough that $\operatorname{supp} p_k \subset V$. Each function

$$\psi_{\dot{v},k}(\delta) = \int \|D\beta(x)\dot{v}\|^{-\delta}\, dp_k^{(\ell)}(\beta)$$

is differentiable and the derivative is given by

$$\psi'_{\dot{v},k}(\delta) = -\int \|D\beta(x)\dot{v}\|^{-\delta} \log\|D\beta(x)\dot{v}\|\, dp_k^{(\ell)}(\beta).$$

Note that $\psi'_{\dot{v},k}(0) = -\int \log\|D\beta(x)\dot{v}\|\, dp_k^{(\ell)}(\beta)$. Since $p_k^{(\ell)} \to p^{(\ell)}$ in the weak* topology, and the function $\beta \mapsto \log\|D\beta(x)\dot{v}\|$ is continuous on V, we get that

$$\lim_k \psi'_{\dot{v},k}(0) = -\int \log\|D\beta(x)\dot{v}\|\, dp^{(\ell)}(\beta) \leq -2c \quad \text{for every } \dot{v}.$$

This convergence is uniform on $\dot{v}$, as the sequence of functions $\dot{v} \mapsto \psi'_{\dot{v},k}(0)$ is equicontinuous. So, assuming that k is large enough, $\psi'_{\dot{v},k}(0) \le -3c/2$ for every $\dot{v}$. Since β and $\dot{v}$ live inside compact sets V and $T^1_x P$, respectively, the factor $\|D\beta(x)\dot{v}\|^{-\delta}$ is uniformly close to 1 if δ is close to 0. Thus, in particular, $\psi'_{\dot{v},k}(\delta) \le -c$ for every $\dot{v}$, every large k and every small δ. Since $\psi_{\dot{v},k}(0) = 1$, this gives the claim. □

Fix δ as in Lemma 10.8 and take k to be large enough that the conclusion of the lemma holds; further conditions will be imposed on k along the way. For notational simplicity, we write $\Pi = p^{(\ell)}$ and $\Pi_k = p_k^{(\ell)}$ for each k.

Let $U_1 \subset U_0$ be an open neighborhood of x such that

$$\int d(\beta(y), \beta(z))^{-\delta} \, d\Pi_k(\beta) \le (1 - c\delta) d(y,z)^{-\delta}, \tag{10.16}$$

for any pair of distinct points $y, z \in U_1$. Since x is a p-invariant point, $\beta(x) = x$ for every $\beta \in \operatorname{supp}\Pi$. Let K be a compact neighborhood of the support of Π and $U_1 \supset U_2 \supset U_3 \supset U_4 \supset U_5$ be open neighborhoods of x such that

$$\beta^{-1}(U_2) \subset U_1 \text{ and } \bar{U}_3 \subset U_2 \text{ and } \bar{U}_4 \subset U_3 \text{ and } \beta(U_5) \subset U_4 \quad \text{for every } \beta \in K.$$

Take k to be large enough that $\operatorname{supp}\Pi_k \subset K$.

The δ-energy of every $\eta_k \mid U_1$ is finite, because the η_k are assumed to be non-atomic. The strategy for the proof of Theorem 10.5 is to use the expansion property (10.16) to find a uniform bound for these δ-energies. That ensures that the δ-energy remains finite at $k = \infty$, contradicting the existence of a fat atom for $\eta \mid U_1$ at x. The main step for realizing this strategy is the following proposition:

Proposition 10.9 *For each $k \ge 1$, let ξ_k be any symmetric self-coupling of $\eta_k \mid U_1$. Then there exists $C > 0$ such that*

$$e_\delta(\eta_k \mid U_1) \le (1 - c\delta) E_\delta(\xi_k) + C \quad \textit{for every } k \textit{ sufficiently large.}$$

Theorem 10.5 can be deduced as follows. Suppose there exists a subsequence $(k_j)_j \to \infty$ such that η_{k_j} has no atoms in U_1 and, in particular, $e_\delta(\eta_{k_j} \mid U_1)$ is finite. Let ξ_{k_j} be a δ-optimal symmetric self-coupling of η_{k_j}; that is, such that $E_\delta(\xi_{k_j}) = e_\delta(\eta_{k_j} \mid U_1)$. Then Proposition 10.9 gives that $e_\delta(\eta_{k_j} \mid U_1) \le C/(c\delta)$. Let $\hat{\eta}$ be any accumulation point of the sequence $(\eta_{k_j} \mid U_1)_j$. On the one hand, by lower semi-continuity,

$$e_\delta(\hat{\eta}) \le \limsup_j e_\delta(\eta_{k_j} \mid U_1) \le \frac{C}{c\delta} < \infty.$$

On the other hand, $\eta \mid U_1 \le \hat{\eta} \le \eta \mid \bar{U}_1$ and, in particular,

$$\hat{\eta}(\{x\}) > \frac{9}{10}\|\hat{\eta}\|.$$

The latter implies that $e_\delta(\hat{\eta}) = \infty$, which contradicts the previous conclusion. This contradiction proves Theorem 10.5.

10.5.1 Proof of Proposition 10.9

We need to construct a (symmetric) self-coupling of $\eta_k \mid U_1$ whose δ-energy is significantly smaller than that of ξ_k. A good starting point is the *diagonal convolution* $\tilde{\xi}_k$ of Π_k and ξ_k; that is, the image of $\Pi_k \times \xi_k$ under the *diagonal push-forward* $(\beta,(y,z)) \mapsto (\beta(y),\beta(z))$. Equivalently,

$$\tilde{\xi}_k(D_1,D_2) = \int \xi_k(\beta^{-1}(D_1) \times \beta^{-1}(D_2))\, d\Pi_k(\beta) \tag{10.17}$$

for any measurable sets $D_1, D_2 \subset \mathbb{PR}^d$. Indeed, it is clear that $\tilde{\xi}_k$ is a symmetric measure and the expansion property (10.16) implies that

$$\begin{aligned} E_\delta(\tilde{\xi}_k) &= \int d(y,z)^{-\delta}\, d\tilde{\xi}_k(y,z) = \int d(\beta(y),\beta(z))^{-\delta}\, d\Pi_k(\beta)\, d\xi_k(y,z) \\ &\le (1-c\delta)\int d(y,z)^{-\delta}\, d\xi_k(y,z) = (1-c\delta)E_\delta(\xi_k). \end{aligned} \tag{10.18}$$

However, the restriction of $\tilde{\xi}_k$ to $U_1 \times U_1$ is *not* a self-coupling of $\eta_k \mid U_1$. Indeed, let η_k^1 be the projection of $\tilde{\xi}_k \mid U_1 \times U_1$ on either coordinate (thus, $\tilde{\xi}_k \mid U_1 \times U_1$ is a symmetric self-coupling of η_k^1). The next lemma shows that $\eta_k^1 \le \eta_k \mid U_1$ and it describes the difference between the two measures precisely. There are two terms, i_k and o_k, that correspond to displacements of mass in opposite directions across the border of U_1 under the diagonal push-forward: i_k accounts for mass that is mapped from the outside into $U_1 \times U_1$ whereas o_k comes from mass that overflows $U_1 \times U_1$. See Figure 10.2.

Lemma 10.10 $(\eta_k \mid U_1) - \eta_k^1 = i_k + o_k$ *where* i_k *is the restriction to* U_1 *of* $\Pi_k * (\eta_k \mid U_1^c)$ *and* o_k *is the projection of* $\tilde{\xi}_k \mid (U_1 \times U_1^c)$ *on the first coordinate. Moreover,*

$$\|o_k\| \le \|i_k\| \quad \textit{and} \quad \limsup_k \|i_k\| \le \frac{1}{10}\eta(\{x\}).$$

Proof It follows from (10.17) that

$$\pi_*^j \tilde{\xi}_k = \int (\eta_k \mid U_1) \circ \beta^{-1}\, d\Pi_k(\beta) = \Pi_k * (\eta_k \mid U_1) \quad \text{for } j = 1,2.$$

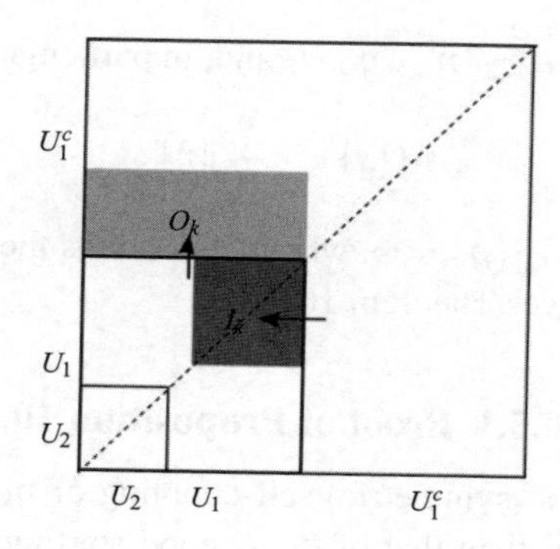

Figure 10.2 Diagonal push-forward: mass in region I_k projects to i_k and mass in region O_k projects to o_k

Combining this with the fact that η_k is Π_k-stationary, we find

$$\pi^1_* \tilde{\xi}_k = \Pi_k * \eta_k - \Pi_k * (\eta_k \mid U_1^c) = \eta_k - \Pi_k * (\eta_k \mid U_1^c),$$

and so $\pi^1_* \tilde{\xi}_k \mid U_1 = (\eta_k \mid U_1) - i_k$. Moreover,

$$\pi^1_* \tilde{\xi}_k \mid U_1 = \pi^1_* (\tilde{\xi}_k \mid U_1 \times U_1) + \pi^1_* (\tilde{\xi}_k \mid U_1 \times U_1^c) = \eta^1_k + o_k.$$

The first claim is a direct consequence of these two equalities.

Next, observe that

$$[\Pi_k * (\eta_k \mid U_1)](U_1^c) + [\Pi_k * (\eta_k \mid U_1)](U_1) = [\Pi_k * (\eta_k \mid U_1)](P) = \eta_k(U_1)$$

and, using that η_k is Π_k-stationary,

$$[\Pi_k * (\eta_k \mid U_1^c)](U_1) + [\Pi_k * (\eta_k \mid U_1)](U_1) = [\Pi_k * \eta_k](U_1) = \eta_k(U_1).$$

This implies that

$$[\Pi_k * (\eta_k \mid U_1^c)](U_1) = [\Pi_k * (\eta_k \mid U_1)](U_1^c). \tag{10.19}$$

The left-hand side is precisely $\|i_k\|$. As for the right-hand side,

$$[\Pi_k * (\eta_k \mid U_1)](U_1^c) = (\pi^2_* \tilde{\xi}_k)(U_1^c) = \tilde{\xi}_k(P \times U_1^c) \geq \tilde{\xi}_k(U_1 \times U_1^c) = \|o_k\|.$$

This proves that $\|o_k\| \leq \|i_k\|$.

Finally, the first part of the lemma implies that $i_k \leq \eta_k \mid U_1$. Moreover, the choice of U_2 ensures that $U_2 \cap \beta^{-1}(U_1) = \emptyset$ for every $\beta \in \operatorname{supp} \Pi_k$ and so $i_k(U_2) = 0$. Thus,

$$\limsup_k \|i_k\| \leq \limsup_k \eta_k(\bar{U}_1 \setminus U_2) \leq \eta(\bar{U}_1 \setminus U_2) \leq \frac{1}{10}\eta(\{x\}).$$

This finishes the proof of the lemma. □

According to Lemma 10.10, the difference $(\eta_k \mid U_1) - \eta_k^1$ is positive and relatively small. This suggests that we try to construct the self-coupling of $\eta_k \mid U_1$ we are looking for, with small δ-energy, by adding suitable correcting terms to the restriction of $\tilde{\xi}_k$ to $U_1 \times U_1$. These correcting terms should be concentrated outside a neighborhood of the diagonal, if possible, so that their contribution to the total energy is bounded. We choose:

$$\begin{aligned}\bar{\xi}_k = &\left[(\tilde{\xi}_k \mid U_1 \times U_1) - \frac{\|\zeta_k\|}{\|\eta_k^4\|}(\tilde{\xi}_k \mid U_4 \times U_4)\right] \\ &+ \frac{1}{\|i_k\|}\left[(o_k \mid U_3) \times i_k + i_k \times (o_k \mid U_3)\right] \\ &+ \frac{1}{\|\eta_k^4\|}\left[(\zeta_k \times \eta_k^4) + (\eta_k^4 \times \zeta_k)\right]\end{aligned} \tag{10.20}$$

where η_k^4 is the projection (on either coordinate) of $\tilde{\xi}_k \mid U_4 \times U_4$ and

$$\zeta_k = \left(1 - \frac{\|o_k \mid U_3\|}{\|i_k\|}\right) i_k + (o_k \mid U_3^c).$$

Lemma 10.11 *Assume that $k \geq 1$ is large enough. Then $\bar{\xi}_k$ is a symmetric self-coupling of $\eta_k \mid U_1$. Moreover,*

$$E_\delta(\bar{\xi}_k) \leq E_\delta(\tilde{\xi}_k \mid U_1 \times U_1) + C,$$

where C is a positive constant independent of k.

Proof It is clear that the three terms on the right-hand side of (10.20) are symmetric. Moreover, $\|o_k \mid U_3\| \leq \|o_k\| \leq \|i_k\|$. This ensures that ζ_k is a positive measure. Now it is clear that the last two terms in (10.20) are positive. To see that the first one is also positive, it suffices to check that $\|\zeta_k\| < \|\eta_k^4\|$. That is easily done as follows. We have $\eta_k(U_5) \geq (9/10)\eta_k(U_1)$ and that implies $\xi_k(U_5 \times U_5) \geq (8/10)\eta_k(U_1)$. Since $\beta^{-1}(U_4) \supset U_5$ for every $\beta \in \operatorname{supp} \Pi_k$, the definition (10.17) gives that $\tilde{\xi}_k(U_4 \times U_4) \geq \xi_k(U_5 \times U_5)$. Thus,

$$\|\eta_k^4\| = \|\tilde{\xi}_k(U_4 \times U_4)\| \geq \frac{8}{10}\eta_k(U_1) \geq \frac{8}{10}\eta(\{x\}).$$

Furthermore, assuming k is sufficiently large,

$$\|\zeta_k\| \leq \|i_k\| + \|o_k\| \leq 2\|i_k\| \leq \frac{3}{10}\eta(\{x\}).$$

In particular, $\|\zeta_k\| < \|\eta_k^4\|$ as we wanted to prove.

Next, observe that the projection of $\bar{\xi}_k$ on either coordinate is equal to

$$\left[\eta_k^1 - \frac{\|\zeta_k\|}{\|\eta_k^4\|}\eta_k^4\right] + \left[(o_k \mid U_3) + \frac{\|o_k \mid U_3\|}{\|i_k\|} i_k\right] + \left[\zeta_k + \frac{\|\zeta_k\|}{\|\eta_k^4\|}\eta_k^4\right]$$

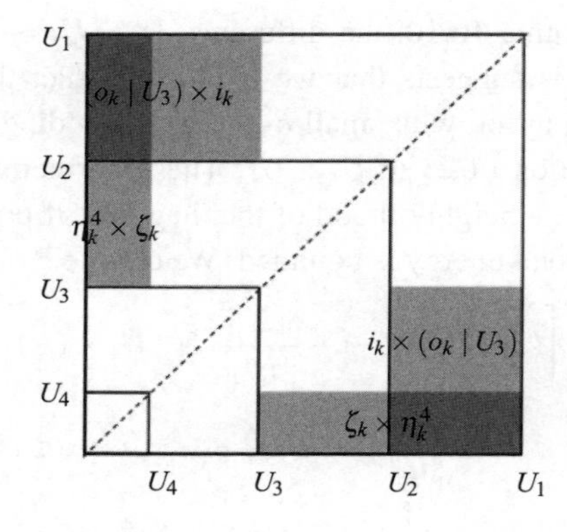

Figure 10.3 Coupling terms far from the diagonal

and that adds up to $\eta_k^1 + i_k + o_k = \eta_k \mid U_1$. So, $\bar{\xi}_k$ is a self-coupling of $\eta_k \mid U_1$ as claimed.

The final step is to estimate the δ-energy of $\bar{\xi}_k$. It is important to notee that the two correcting terms

$$\frac{1}{\|i_k\|}\left[(o_k \mid U_3) \times i_k + i_k \times (o_k \mid U_3)\right] \quad \text{and} \quad \frac{1}{\|\eta_k^4\|}\left[(\zeta_k \times \eta_k^4) + (\eta_k^4 \times \zeta_k)\right]$$

are supported away from the diagonal of $P \times P$. Indeed, $o_k \mid U_3$ is concentrated on U_3 while i_k is concentrated on U_2^c, as we have seen. Similarly, ζ_k is concentrated on U_3^c while η_k^4 is concentrated on U_4. See Figure 10.3. Moreover, their masses are uniformly bounded (by 1, say). Thus, the δ-energy of the sum of these two terms is bounded by some constant $C > 0$, for every large k. As for the first term in (10.20), it is clear that its δ-energy is bounded by $E_\delta(\tilde{\xi}_k \mid U_1 \times U_1)$. Thus, $E_\delta(\bar{\xi}_k) \le E_\delta(\tilde{\xi}_k \mid U_1 \times U_1) + C$, as claimed. □

Combining (10.18) with Lemma 10.11, we get

$$e_\delta(\eta_k^i) \le E_\delta(\bar{\xi}_k) \le E_\delta(\tilde{\xi}_k \mid U_1 \times U_1) + C \le E_\delta(\tilde{\xi}_k) + C \le (1 - c\delta) E_\delta(\xi_k) + C$$

as claimed in Proposition 10.9.

10.6 Final comments

We saw in Chapter 9 that discontinuity of the Lyapunov exponents is common among cocycles with low regularity: continuous or Hölder-continuous cocycles with small Hölder constant. On the other hand, Theorems 1.3 and 10.1 show that, at least in dimension 2, for locally constant cocycles (cocycles with "infinite Hölder constant") one always has continuity. In what follows, we present a possible global scenario for 2-dimensional cocycles over "chaotic" systems, encompassing these two contrasting conclusions.

Let $X = \{1, \dots, m\}$ and $f : M \to M$ be the shift map on $M = X^{\mathbb{Z}}$. Fix $\theta \in (0,1)$ and endow M with the metric

$$d_\theta(x,y) = \theta^{N(x,y)}, \text{ where } N(x,y) = \max\{N \geq 0 : x_n = y_n \text{ for all } |n| < N\}.$$

Let $A : M \to \mathrm{GL}(2)$ be r-Hölder continuous for this metric. In other words, there exists $C_1 > 0$ such that

$$\|A(x) - A(y)\| \leq C_1 \theta^{Nr} \quad \text{for any } x, y \text{ with } x_n = y_n \text{ for all } |n| < N. \tag{10.21}$$

The cocycle F defined by A over f is *fiber-bunched* if there exists $C_2 > 0$ and $\lambda < 1$ such that

$$\|A^n(x)\| \, \|A^n(x)^{-1}\| \, \theta^{nr} \leq C_2 \lambda^n \text{ for every } x \in M \text{ and } n \geq 1. \tag{10.22}$$

This notion was introduced in [37], under a different name.

Conjecture 10.12 *Lyapunov exponents vary continuously restricted to the subset of fiber-bunched elements $A : M \to \mathrm{GL}(2)$ of the space of r-Hölder-continuous cocycles.*

Observe that in Section 9.3 we had $\|A\| = \|A^{-1}\| = \sigma$ and $\theta = 1/2$, and so the cocycle is fiber-bunched if and only if $\sigma^2 < 2^r$. Thus, the hypothesis $\sigma > 2^{2r}$ in Theorem 9.22 is incompatible with fiber-bunching.

It is clear from the definition that fiber-bunched cocycles form an open subset of the space of r-Hölder-continuous cocycles for any $r > 0$. An important feature of fiber-bunched cocycles is that they admit invariant stable and unstable holonomies. Let us explain this.

The *local stable set* $W^{\mathrm{s}}_{\mathrm{loc}}(x)$ of a point $x \in M$ is the set of all $y \in M$ that have the same positive part; that is, such that $\pi^+(x) = \pi^+(y)$. The *local unstable set* is defined analogously, requiring $\pi^-(x) = \pi^-(y)$ instead. The *stable and unstable sets* are defined by

$$W^{\mathrm{s}}(x) = \bigcup_{n=0}^{\infty} f^{-n}\left(W^{\mathrm{s}}_{\mathrm{loc}}(f^n(x))\right) \quad \text{and} \quad W^{\mathrm{u}}(x) = \bigcup_{n=0}^{\infty} f^{n}\left(W^{\mathrm{u}}_{\mathrm{loc}}(f^{-n}(x))\right),$$

respectively. An *s-holonomy* for the cocycle associated with A is a family of linear maps $h^{\mathrm{s}}_{x,y} : \mathbb{R}^d \to \mathbb{R}^d$, defined for every x and y in the same stable set, such that

(1) $h^{\mathrm{s}}_{y,z} \circ h^{\mathrm{s}}_{x,y} = h^{\mathrm{s}}_{x,z}$ and $h^{\mathrm{s}}_{x,x} = \mathrm{id}$ for every x, y, z in the same stable set;
(2) $A(y) \circ h^{\mathrm{s}}_{x,y} = h^{\mathrm{s}}_{f(x),f(y)} \circ A(x)$ for every x, y in the same stable set;

(3) $(x,y,v) \mapsto h^{\mathrm{s}}_{x,y}(v)$ is continuous, when x and y vary in the set of points in the same local stable set;
(4) there are $C > 0$ and $t > 0$ such that $h^{\mathrm{s}}_{x,y}$ is (C,t)-Hölder for every x and y in the same local stable set.

Analogously, one defines *u-holonomy* for the cocycle. If A is fiber-bunched then

$$h^{\mathrm{s}}_{x,y} = \lim_n A^n(y)^{-1}A^n(x) \quad \text{and} \quad h^{\mathrm{u}}_{x,y} = \lim_n A^{-n}(y)^{-1}A^{-n}(x)$$

are well defined and they yield an s-holonomy and a u-holonomy for the cocycle.

Fiber-bunching is not necessary for the existence of invariant holonomies, of course. For instance, for locally constant cocycles one may always take

$$h^{\mathrm{s}}_{x,y} = \mathrm{id} \text{ for all } y \in W^{\mathrm{s}}_{\mathrm{loc}}(x) \quad \text{and} \quad h^{\mathrm{u}}_{x,y} = \mathrm{id} \text{ for all } y \in W^{\mathrm{u}}_{\mathrm{loc}}(x).$$

Thus, the following statement contains both Theorem 1.3 and Conjecture 10.12:

Conjecture 10.13 *Let $\mathcal{H}$ be a family of r-Hölder-continuous $A : M \to \mathrm{GL}(2)$ such that stable and unstable holonomies, $h^{\mathrm{s}}_{A,x,y}$ and $h^{\mathrm{u}}_{A,x,y}$, exist for all $A \in \mathcal{H}$ and vary continuously on $\mathcal{H}$. Then the Lyapunov exponents vary continuously on $\mathcal{H}$.*

Here, the continuity condition means that $(x,y,v,A) \mapsto h^{\mathrm{s}}_{A,x,y}(v)$ is continuous when x and y vary in the set of points in the same local stable set, and $(x,y,v,A) \mapsto h^{\mathrm{u}}_{A,x,y}(v)$ is continuous when x and y vary in the set of points in the same local unstable set. Moreover, the Hölder constants C and t in condition (4) are assumed to be locally uniform in A.

Hyperbolic cocycles need not be fiber-bunched neither do the cocycles with $\lambda_- = \lambda_+$. Thus, Corollary 9.3 and Lemma 9.4 imply that the converse to Conjecture 10.12 is not true in general. Instead, we state:

Conjecture 10.14 *Consider any r-continuous function $A : M \to \mathrm{GL}(2)$ such that $\lambda_-(A,\mu) < \lambda_+(A,\mu)$ and the cocycle defined by A over f is not hyperbolic. If A is not fiber-bunched then there exists a sequence $(A_n)_n$ converging to A in the space of r-Hölder-continuous functions such that* $\lim_n \lambda_+(A_n,\mu) < \lambda_+(A,\mu)$.

Young [118] constructed a C^1-open set of $\mathrm{SL}(2)$-cocycles over an expanding circle map which are neither hyperbolic nor fiber-bunched and whose Lyapunov exponents are uniformly bounded from zero. The estimates are not sufficient to decide whether the Lyapunov exponents vary continuously in this set. Still, if such an open set can be constructed for Hölder-continuous cocycles, this could be a good test case for Conjecture 10.14.

10.7 Notes

Theorem 10.1 was proven by Bocker and Viana [35], based on certain estimates of the random walk induced by the cocycle on $\mathbb{PR}^2$, which they obtained through a suitable discretization of the projective space. Avila, Eskin and Viana [12] use a very different analysis of the random walk to extend the theorem to arbitrary dimension $d \geq 2$. The presentation in Sections 10.1 through 10.5 is based on the latter.

The dependence of Lyapunov exponents on the linear cocycle or the base dynamics has been studied by many authors. Ruelle [104] proved real-analytic dependence of the largest exponent on the cocycle, for linear cocycles admitting an invariant convex cone field. Short afterwards, Furstenberg and Kifer [57, 70] and Hennion [61] studied the dependence of the largest exponent of random matrices on the probability distribution, proving continuity with respect to the weak* topology in the almost irreducible case; that is, when there is at most one invariant subspace. Kifer [70] observed that Lyapunov exponents may jump when the probability vector degenerates (Exercise 1.8) and Johnson [67] also found examples of discontinuous dependence of the exponent on the energy E, for Schrödinger cocycles over quasi-periodic flows.

For random matrices satisfying strong irreducibility and the contraction property, Le Page [94, 95] proved local Hölder continuity and even smooth dependence of the largest exponent on the cocycle; the assumptions ensure that the largest exponent is simple (multiplicity 1), by work of Guivarc'h and Raugi [60] and Gol'dsheid and Margulis [58]. Le Page's result cannot be improved: a construction of Halperin (see Simon and Taylor [108]) shows that for every $\alpha > 0$ one can find random Schrödinger cocycles near which the exponents fail to be α-Hölder continuous. For random matrices with finitely many values and, more generally, for locally constant cocycles over Markov shifts, Peres [98] showed that simple exponents are locally real-analytic functions of the transition data. Recently, Bourgain and Jitomirskaya [40, 41] proved continuous dependence of the exponents on the energy E, for quasi-periodic Schrödinger cocycles.

10.8 Exercises

Exercise 10.1 Prove that if $(p_k)_k$ converges to p in the space $\mathcal{G}(2)$ then $(\lambda_+(p_k) + \lambda_-(p_k))_k$ converges to $\lambda_+(p) + \lambda_-(p)$. Conclude that $\lambda_+(p_k) \to \lambda_+(p)$ if and only if $\lambda_-(p_k) \to \lambda_-(p)$.

Exercise 10.2 Prove the following properties of the *convolution* operations:

(1) $* : \mathscr{G}(d) \times \mathscr{P}(d) \to \mathscr{P}(d)$ defined in (10.3) is continuous and it satisfies $(p_1 * p_2) * \eta = p_1 * (p_2 * \eta)$.
(2) $* : \mathscr{G}(d) \times \mathscr{G}(d) \to \mathscr{G}(d)$ defined in (10.4) is commutative, associative and continuous. Is there a unit element?

Exercise 10.3 Verify the equality (10.5) and prove that

$$\frac{\|\dot{v}\|}{\|\alpha\| \, \|\alpha^{-1}\|} \leq \|D\alpha([v])\dot{v}\| \leq \|\alpha\| \, \|\alpha^{-1}\| \, \|\dot{v}\|$$

for every $[v] \in \mathbb{PR}^d$ and $\dot{v} \in \{v\}^{\perp}$.

Exercise 10.4 Prove that if $(p_k)_k$ converges to p in $\mathscr{G}(d)$ then $(\operatorname{supp} p_k)_k$ converges to $\operatorname{supp} p$ relative to the *Hausdorff distance*, defined in the space of compact subsets of $\mathrm{GL}(d)$ by

$$d_H(K_1, K_2) = \inf\{r > 0 : K_1 \subset B_r(K_2) \text{ and } K_2 \subset B_r(K_1)\}.$$

Exercise 10.5 Prove the following.

(1) The function $\xi \mapsto E_\delta(\xi)$ is lower semi-continuous with respect to the weak* topology in the space of measures with a given mass on $N \times N$.
(2) The function $\eta \mapsto e_\delta(\eta)$ is lower semi-continuous with respect to the weak* topology in the space of measures with a given mass on N.

Exercise 10.6 Let η be a probability measure on a compact metric space P. Check that the function $U \mapsto e_\delta(\eta \mid U)$ may not be monotone.

Exercise 10.7 Let η be a Borel measure on some compact metric space P such that $\eta(\{x\}) < \|\eta\|/2$ for every $x \in P$. Show that there exists $\tau > 0$ such that $\eta(B(z,\tau)) < \eta(B(z,\tau)^c)$ for every $z \in P$.

Exercise 10.8 Let η be a probability on $[0,1]$ of the form $\eta = (\delta_0 + \mu)/2$, where μ is a probability on $(0,1]$. Check that, depending on μ, the δ-energy of η may be finite or infinite.

Exercise 10.9 State and prove an analogue of Lemma 10.7 for couplings of any two measures η_1 and η_2 with the same mass.

Exercise 10.10 Check that the fiber-bunching property (10.22) does not depend on the choice of θ in the definition of the distance.

References

[1] P. Anderson. Absence of diffusion in certain random lattices. *Phys. Rev.*, 109:1492–1505, 1958.

[2] D. V. Anosov. Geodesic flows on closed Riemannian manifolds of negative curvature. *Proc. Steklov Math. Inst.*, 90:1–235, 1967.

[3] V. Araújo and M. Bessa. Dominated splitting and zero volume for incompressible three-flows. *Nonlinearity*, 21:1637–1653, 2008.

[4] A. Arbieto and J. Bochi. L^p-generic cocycles have one-point Lyapunov spectrum. *Stoch. Dyn.*, 3:73–81, 2003.

[5] L. Arnold. *Random Dynamical Systems*. Springer-Verlag, 1998.

[6] L. Arnold and N. D. Cong. On the simplicity of the Lyapunov spectrum of products of random matrices. *Ergod. Theory Dynam. Sys.*, 17:1005–1025, 1997.

[7] L. Arnold and V. Wishtutz (Eds.). *Lyapunov Exponents (Bremen, 1984)*, volume 1186 of *Lecture Notes in Math.* Springer, 1986.

[8] A. Avila. Density of positive Lyapunov exponents for quasiperiodic $\mathrm{SL}(2,\mathbb{R})$-cocycles in arbitrary dimension. *J. Mod. Dyn.*, 3:631–636, 2009.

[9] A. Avila and J. Bochi. Lyapunov exponents: parts I & II. Notes of mini-course given at the School on Dynamical Systems, ICTP, 2008.

[10] A. Avila and J. Bochi. Proof of the subadditive ergodic theorem. Preprint www.mat.puc-rio.br/∼jairo/.

[11] A. Avila and J. Bochi. A formula with some applications to the theory of Lyapunov exponents. *Israel J. Math.*, 131:125–137, 2002.

[12] A. Avila, A .Eskin and M. Viana. Continuity of Lyapunov exponents of random matrix products. In preparation.

[13] A. Avila, J. Santamaria, and M. Viana. Holonomy invariance: rough regularity and Lyapunov exponents. *Astérisque*, 358:13–74, 2013.

[14] A. Avila and M. Viana. Simplicity of Lyapunov spectra: a sufficient criterion. *Port. Math.*, 64:311–376, 2007.

[15] A. Avila and M. Viana. Simplicity of Lyapunov spectra: proof of the Zorich-Kontsevich conjecture. *Acta Math.*, 198:1–56, 2007.

[16] A. Avila and M. Viana. Extremal Lyapunov exponents: an invariance principle and applications. *Inventiones Math.*, 181:115–178, 2010.

[17] A. Avila, M. Viana, and A. Wilkinson. Absolute continuity, Lyapunov exponents and rigidity I: geodesic flows. Preprint www.preprint.impa.br 2011.

[18] L. Barreira and Ya. Pesin. *Nonuniform Hyperbolicity: Dynamics of Systems with Nonzero Lyapunov Exponents*. Cambridge University Press, 2007.

[19] L. Barreira, Ya. Pesin and J. Schmeling. Dimension and product structure of hyperbolic measures. *Ann. of Math.*, 149:755–783, 1999.

[20] M. Bessa. Dynamics of generic 2-dimensional linear differential systems. *J. Differential Equations*, 228:685–706, 2006.

[21] M. Bessa. The Lyapunov exponents of generic zero divergence three-dimensional vector fields. *Ergodic Theory Dynam. Systems*, 27:1445–1472, 2007.

[22] M. Bessa. Dynamics of generic multidimensional linear differential systems. *Adv. Nonlinear Stud.*, 8:191–211, 2008.

[23] M. Bessa and J. L. Dias. Generic dynamics of 4-dimensional C^2 Hamiltonian systems. *Comm. Math. Phys.*, 281:597–619, 2008.

[24] M. Bessa and J. Rocha. Contributions to the geometric and ergodic theory of conservative flows. *Ergodic Theory Dynam. Sys.*, 33:1709–1731, 2013.

[25] M. Bessa and H. Vilarinho. Fine properties of lp-cocycles. *J. Differential Equations*, 256:2337–2367, 2014.

[26] G. Birkhoff. *Lattice Theory*, volume 25. A.M.S. Colloq. Publ., 1967.

[27] J. Bochi. Discontinuity of the Lyapunov exponents for non-hyperbolic cocycles. Preprint www.mat.puc-rio.br/~jairo/.

[28] J. Bochi. The multiplicative ergodic theorem of Oseledets. Preprint www.mat.puc-rio.br/~jairo/.

[29] J. Bochi. Proof of the Oseledets theorem in dimension 2 via hyperbolic geometry. Preprint www.mat.puc-rio.br/~jairo/.

[30] J. Bochi. Genericity of zero Lyapunov exponents. *Ergod. Theory Dynam. Sys.*, 22:1667–1696, 2002.

[31] J. Bochi. C^1-generic symplectic diffeomorphisms: partial hyperbolicity and zero centre Lyapunov exponents. *J. Inst. Math. Jussieu*, 8:49–93, 2009.

[32] J. Bochi and M. Viana. Pisa lectures on Lyapunov exponents. In *Dynamical Systems. Part II*, Pubbl. Cent. Ric. Mat. Ennio Giorgi, pages 23–47. Scuola Norm. Sup., 2003.

[33] J. Bochi and M. Viana. Lyapunov exponents: how frequently are dynamical systems hyperbolic? In *Modern Dynamical Systems and Applications*, pages 271–297. Cambridge University Press, 2004.

[34] J. Bochi and M. Viana. The Lyapunov exponents of generic volume-preserving and symplectic maps. *Ann. of Math.*, 161:1423–1485, 2005.

[35] C. Bocker and M. Viana. Continuity of Lyapunov exponents for 2D random matrices. Preprint www.impa.br/~viana/out/bernoulli.pdf.

[36] C. Bonatti, L. J. Díaz, and M. Viana. *Dynamics Beyond Uniform Hyperbolicity*, volume 102 of *Encyclopaedia of Mathematical Sciences*. Springer-Verlag, 2005.

[37] C. Bonatti, X. Gómez-Mont, and M. Viana. Généricité d'exposants de Lyapunov non-nuls pour des produits déterministes de matrices. *Ann. Inst. H. Poincaré Anal. Non Linéaire*, 20:579–624, 2003.

[38] C. Bonatti and M. Viana. SRB measures for partially hyperbolic systems whose central direction is mostly contracting. *Israel J. Math.*, 115:157–193, 2000.

[39] N. Bourbaki. *Algebra. I. Chapters 1–3*. Elements of Mathematics (Berlin). Springer-Verlag, 1989. Translated from the French, Reprint of the 1974 edition.

[40] J. Bourgain. Positivity and continuity of the Lyapounov exponent for shifts on $\mathbb{T}^d$ with arbitrary frequency vector and real analytic potential. *J. Anal. Math.*, 96:313–355, 2005.

[41] J. Bourgain and S. Jitomirskaya. Continuity of the Lyapunov exponent for quasiperiodic operators with analytic potential. *J. Statist. Phys.*, 108:1203–1218, 2002.

[42] M. Cambrainha. *Generic symplectic cocycles are hyperbolic*. PhD thesis, IMPA, 2013.

[43] C. Castaing and M. Valadier. *Convex Analysis and Measurable Multifunctions*. Lecture Notes in Mathematics, Vol. 580. Springer-Verlag, 1977.

[44] J. E. Cohen, H. Kesten, and C. M. Newman (Eds.). *Random Matrices and their Applications (Brunswick, Maine, 1984)*, volume 50 of *Contemp. Math.* Amer. Math. Soc., 1986.

[45] J. Conway. *Functions of One Complex Variable. II*, volume 159 of *Graduate Texts in Mathematics*. Springer-Verlag, 1995.

[46] A. Crisanti, G.Paladin, and A. Vulpiani. *Products of Random Matrices in Statistical Physics*, volume 104 of *Springer Series in Solid-State Sciences*. Springer-Verlag, 1993. With a foreword by Giorgio Parisi.

[47] D. Damanik. Schrödinger operators with dynamically defined potentials. In preparation.

[48] David Damanik. Lyapunov exponents and spectral analysis of ergodic Schrödinger operators: a survey of Kotani theory and its applications. In *Spectral Theory and Mathematical Physics: a Festschrift in Honor of Barry Simon's 60th birthday*, volume 76 of *Proc. Sympos. Pure Math.*, pages 539–563. Amer. Math. Soc., 2007.

[49] L. J. Díaz and R. Ures. Persistent homoclinic tangencies at the unfolding of cycles. *Ann. Inst. H. Poincaré, Anal. Non-linéaire*, 11:643–659, 1996.

[50] A. Douady and J. C. Earle. Conformally natural extension of homeomorphisms of the circle. *Acta Math.*, 157:23–48, 1986.

[51] R. Dudley, H. Kunita, F. Ledrappier, and P. Hennequin (Eds.). *École d'été de probabilités de Saint-Flour, XII—1982*, volume 1097 of *Lecture Notes in Math.* Springer, 1984.

[52] J. Franks. Anosov diffeomorphisms. In *Global Analysis (Proc. Sympos. Pure Math., Vol. XIV, Berkeley, Calif., 1968)*, pages 61–93. Amer. Math. Soc., 1970.

[53] N. Friedman. *Introduction to Ergodic Theory*. Van Nostrand, 1969.

[54] H. Furstenberg. Non-commuting random products. *Trans. Amer. Math. Soc.*, 108:377–428, 1963.

[55] H. Furstenberg. Boundary theory and stochastic processes on homogeneous spaces. In *Harmonic Analysis in Homogeneous Spaces*, volume XXVI of *Proc. Sympos. Pure Math. (Williamstown MA, 1972)*, pages 193–229. Amer. Math. Soc., 1973.

[56] H. Furstenberg and H. Kesten. Products of random matrices. *Ann. Math. Statist.*, 31:457–469, 1960.

[57] H. Furstenberg and Yu. Kifer. Random matrix products and measures in projective spaces. *Israel J. Math*, 10:12–32, 1983.

[58] I. Ya. Gol'dsheid and G. A. Margulis. Lyapunov indices of a product of random matrices. *Uspekhi Mat. Nauk.*, 44:13–60, 1989.

[59] Y. Guivarc'h. Marches aléatories à pas markovien. *Comptes Rendus Acad. Sci. Paris*, 289:211–213, 1979.

[60] Y. Guivarc'h and A. Raugi. Products of random matrices: convergence theorems. *Contemp. Math.*, 50:31–54, 1986.

[61] H. Hennion. Loi des grands nombres et perturbations pour des produits réductibles de matrices aléatoires indépendantes. *Z. Wahrsch. Verw. Gebiete*, 67:265–278, 1984.

[62] M. Herman. Une méthode nouvelle pour minorer les exposants de Lyapunov et quelques exemples montrant le caractère local d'un théorème d'Arnold et de Moser sur le tore de dimension 2. *Comment. Math. Helvetici*, 58:453–502, 1983.

[63] A. Herrera. *Simplicity of the Lyapunov spectrum for multidimensional continued fraction algorithms*. PhD thesis, IMPA, 2009.

[64] F. Rodriguez Hertz, M. A. Rodriguez Hertz, A. Tahzibi, and R. Ures. Maximizing measures for partially hyperbolic systems with compact center leaves. *Ergodic Theory Dynam. Sys.*, 32:825–839, 2012.

[65] E. Hopf. Statistik der geodätischen Linien in Mannigfaltigkeiten negativer Krümmung. *Ber. Verh. Sächs. Akad. Wiss. Leipzig*, 91:261–304, 1939.

[66] S. Jitomirskaya and C. Marx. Dynamics and spectral theory of quasi-periodic Schrödinger type operators. In preparation.

[67] R. Johnson. Lyapounov numbers for the almost periodic Schrödinger equation. *Illinois J. Math.*, 28:397–419, 1984.

[68] A. Katok. Lyapunov exponents, entropy and periodic points of diffeomorphisms. *Publ. Math. IHES*, 51:137–173, 1980.

[69] Y. Katznelson and B. Weiss. A simple proof of some ergodic theorems. *Israel J. Math.*, 42:291–296, 1982.

[70] Yu. Kifer. Perturbations of random matrix products. *Z. Wahrsch. Verw. Gebiete*, 61:83–95, 1982.

[71] Yu. Kifer. *Ergodic Theory of Random Perturbations*. Birkhäuser, 1986.

[72] Yu. Kifer. General random perturbations of hyperbolic and expanding transformations. *J. Analyse Math.*, 47:11–150, 1986.

[73] Yu. Kifer and E. Slud. Perturbations of random matrix products in a reducible case. *Ergodic Theory Dynam. Sys.*, 2:367–382 (1983), 1982.

[74] J. Kingman. The ergodic theorem of subadditive stochastic processes. *J. Royal Statist. Soc.*, 30:499–510, 1968.

[75] O. Knill. The upper Lyapunov exponent of $\mathrm{SL}(2,\mathbf{R})$ cocycles: discontinuity and the problem of positivity. In *Lyapunov Exponents (Oberwolfach, 1990)*, volume 1486 of *Lecture Notes in Math.*, pages 86–97. Springer-Verlag, 1991.

[76] O. Knill. Positive Lyapunov exponents for a dense set of bounded measurable $SL(2,\mathbf{R})$-cocycles. *Ergod. Theory Dynam. Sys.*, 12:319–331, 1992.

[77] S. Kotani. Lyapunov indices determine absolutely continuous spectra of stationary random one-dimensional Schrödinger operators. In *Stochastic Analysis*, pages 225–248. North Holland, 1984.

[78] H. Kunz and Bernard B. Souillard. Sur le spectre des opérateurs aux différences finies aléatoires. *Comm. Math. Phys.*, 78:201–246, 1980/81.

[79] F. Ledrappier. Propriétés ergodiques des mesures de Sinaï *Publ. Math. I.H.E.S.*, 59:163–188, 1984.

[80] F. Ledrappier. Quelques propriétés des exposants caractéristiques. volume 1097 of *Lect. Notes in Math.*, pages 305–396, Springer-Verlag, 1984.

[81] F. Ledrappier. Positivity of the exponent for stationary sequences of matrices. In *Lyapunov Exponents (Bremen, 1984)*, volume 1186 of *Lect. Notes in Math.*, pages 56–73. Springer-Verlag, 1986.

[82] F. Ledrappier and G. Royer. Croissance exponentielle de certain produits aléatorires de matrices. *Comptes Rendus Acad. Sci. Paris*, 290:513–514, 1980.

[83] F. Ledrappier and L.-S. Young The metric entropy of diffeomorphisms. I. Characterization of measures satisfying Pesin's entropy formula. *Ann. of Math.*, 122:509–539, 1985.

[84] F. Ledrappier and L.-S. Young The metric entropy of diffeomorphisms. II. Relations between entropy, exponents and dimension. *Ann. of Math.*, 122:540–574, 1985.

[85] A. M. Lyapunov. *The General Problem of the Stability of Motion*. Taylor & Francis Ltd., 1992. Translated from Edouard Davaux's French translation (1907) of the 1892 Russian original and edited by A.T. Fuller, with an introduction and preface by Fuller, a biography of Lyapunov by V.I. Smirnov, and a bibliography of Lyapunov's works compiled by J.F. Barrett, Lyapunov centenary issue, Reprint of *Internat. J. Control* **55** (1992), no. 3 [MR1154209 (93e:01035)], With a foreword by Ian Stewart.

[86] R. Mañé. A proof of Pesin's formula. *Ergod. Theory Dynam. Sys.*, 1:95–101, 1981.

[87] R. Mañé. Lyapunov exponents and stable manifolds for compact transformations. In *Geometric Dynamics*, volume 1007 of *Lect. Notes in Math.*, pages 522–577. Springer-Verlag, 1982.

[88] R. Mañé. Oseledec's theorem from the generic viewpoint. In *Procs. International Congress of Mathematicians, Vol. 1, 2 (Warsaw, 1983)*, pages 1269–1276, Warsaw, 1984. PWN Publ.

[89] R. Mañé. *Ergodic Theory and Differentiable Dynamics*. Springer-Verlag, 1987.

[90] A. Manning. There are no new Anosov diffeomorphisms on tori. *Amer. J. Math.*, 96:422–429, 1974.

[91] S. Newhouse. On codimension one Anosov diffeomorphisms. *Amer. J. Math.*, 92:761–770, 1970.

[92] V. I. Oseledets. A multiplicative ergodic theorem: Lyapunov characteristic numbers for dynamical systems. *Trans. Moscow Math. Soc.*, 19:197–231, 1968.

[93] J. C. Oxtoby and S. M. Ulam. Measure-preserving homeomorphisms and metrical transitivity. *Ann. of Math.*, 42:874–920, 1941.

[94] É. Le Page. Théorèmes limites pour les produits de matrices aléatoires. In *Probability Measures on Groups (Oberwolfach, 1981)*, volume 928 of *Lecture Notes in Math.*, pages 258–303. Springer-Verlag, 1982.

[95] É. Le Page. Régularité du plus grand exposant caractéristique des produits de matrices aléatoires indépendantes et applications. *Ann. Inst. H. Poincaré Probab. Statist.*, 25:109–142, 1989.

[96] J. Palis and S. Smale. Structural stability theorems. In *Global Analysis*, volume XIV of *Proc. Sympos. Pure Math. (Berkeley 1968)*, pages 223–232. Amer. Math. Soc., 1970.

[97] L. Pastur. Spectral properties of disordered systems in the one-body approximation. *Comm. Math. Phys.*, 75:179–196, 1980.

[98] Y. Peres. Analytic dependence of Lyapunov exponents on transition probabilities. In *Lyapunov Exponents (Oberwolfach, 1990)*, volume 1486 of *Lecture Notes in Math.*, pages 64–80. Springer-Verlag, 1991.

[99] Ya. B. Pesin. Families of invariant manifolds corresponding to non-zero characteristic exponents. *Math. USSR. Izv.*, 10:1261–1302, 1976.

[100] Ya. B. Pesin. Characteristic Lyapunov exponents and smooth ergodic theory. *Russian Math. Surveys*, 324:55–114, 1977.

[101] M. S. Raghunathan. A proof of Oseledec's multiplicative ergodic theorem. *Israel J. Math.*, 32:356–362, 1979.

[102] V. A. Rokhlin. On the fundamental ideas of measure theory. *A. M. S. Transl.*, 10:1–54, 1962. Transl. from Mat. Sbornik 25 (1949), 107–150. First published by the A. M. S. in 1952 as Translation Number 71.

[103] G. Royer. Croissance exponentielle de produits markoviens de matrices. *Ann. Inst. H. Poincaré*, 16:49–62, 1980.

[104] D. Ruelle. Analyticity properties of the characteristic exponents of random matrix products. *Adv. in Math.*, 32:68–80, 1979.

[105] D. Ruelle. Ergodic theory of differentiable dynamical systems. *Inst. Hautes Études Sci. Publ. Math.*, 50:27–58, 1979.

[106] D. Ruelle. Characteristic exponents and invariant manifolds in Hilbert space. *Annals of Math.*, 115:243–290, 1982.

[107] M. Shub. *Global Stability of Dynamical Systems*. Springer-Verlag, 1987.

[108] B. Simon and M. Taylor. Harmonic analysis on $\mathrm{SL}(2,\mathbf{R})$ and smoothness of the density of states in the one-dimensional Anderson model. *Comm. Math. Phys.*, 101:1–19, 1985.

[109] S. Smale. Differentiable dynamical systems. *Bull. Am. Math. Soc.*, 73:747–817, 1967.

[110] A. Tahzibi. Stably ergodic diffeomorphisms which are not partially hyperbolic. *Israel J. Math.*, 142:315–344, 2004.

[111] Ph. Thieullen. Ergodic reduction of random products of two-by-two matrices. *J. Anal. Math.*, 73:19–64, 1997.

[112] M. Viana. Lyapunov exponents and strange attractors. In J.-P. Françoise, G. L. Naber, and S. T. Tsou, editors, *Encyclopedia of Mathematical Physics*. Elsevier, 2006.

[113] M. Viana. Almost all cocycles over any hyperbolic system have nonvanishing Lyapunov exponents. *Ann. of Math.*, 167:643–680, 2008.

[114] M. Viana and K. Oliveira. *Fundamentos da Teoria Ergódica*. Coleção Fronteiras da Matemática. Sociedade Brasileira de Matemática, 2014.

[115] M. Viana and J. Yang. Physical measures and absolute continuity for one-dimensional center directions. *Ann. Inst. H. Poincaré Anal. Non Linéaire*, 30:845–877, 2013.

[116] A. Virtser. On products of random matrices and operators. *Th. Prob. Appl.*, 34:367–377, 1979.

[117] P. Walters. A dynamical proof of the multiplicative ergodic theorem. *Trans. Amer. Math. Soc.*, 335:245–257, 1993.

[118] L.-S. Young. Some open sets of nonuniformly hyperbolic cocycles. *Ergod. Theory Dynam. Sys.*, 13(2):409–415, 1993.

[119] L.-S. Young. Ergodic theory of differentiable dynamical systems. In *Real and Complex Dynamical Systems*, volume NATO ASI Series, C-464, pages 293–336. Kluwer Acad. Publ., 1995.

[120] A. C. Zaanen. *Integration*. North-Holland Publishing Co., 1967. Completely revised edition of *An Introduction to the Theory of Integration*.

Index